# THE
# LEAKEYS

A BIOGRAPHY

# THE
# LEAKEYS

*Mary Bowman-Kruhm*

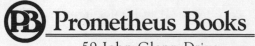

**Prometheus Books**

59 John Glenn Drive
Amherst, New York 14228–2119

Published 2010 by Prometheus Books

Inquiries should be addressed to
Prometheus Books
59 John Glenn Drive
Amherst, New York 14228–2119
VOICE: 716–691–0133
FAX: 716–691–0137
WWW.PROMETHEUSBOOKS.COM

14 13 12 11 10    5 4 3 2 1

Library of Congress Cataloging-in-Publication Data

Bowman-Kruhm, Mary.
   The Leakeys : a biography / Mary Bowman-Kruhm.
      p.  cm.
   Includes bibliographical references and index.
   ISBN 978–1–59102–761–4 (pbk. : alk. paper)
   1. Leakey, L. S. B. (Louis Seymour Bazett), 1903–1972. 2. Leakey, Mary D. (Mary Douglas), 1913–1996. 3. Leakey, Richard E. 4. Paleoanthropologists—Tanzania—Olduvai Gorge—Biography. 5. Physical anthropologists—Tanzania—Olduvai Gorge— Antiquities. 6. Stone age—Tanzania—Olduvai Gorge. 7. Fossil hominids—Tanzania— Olduvai Gorge. 8. Olduvai Gorge (Tanzania)—Antiquities.  I. Title.

GN50.5.B68 2009
569.9092'2—dc22
[B]                                                                            2009024997

Printed in the United States of America

Dedicated to my grandson,
Mike McGonigle, whose company
I enjoy, whether watching him
play soccer or visiting the tortoise
house at Koobi Fora, and to Dr.
Louise Leakey and her team of
young Kenyans who are trying to
find answers to the question of
how we came to be.

# CONTENTS

8 **CONTENTS**

# ACKNOWLEDGMENTS

Special thanks for their thoughtful review of the manuscript to Dr. Sara Patricia Chavarria, Curriculum Development and Consultation Services, Tucson, AZ; Dr. Cathy Willermet, Adjunct Faculty, Sociology and Anthropology, University of Texas at El Paso; Dr. Thomas Gundling, Assistant Professor of Anthropology, William Paterson University; Dr. Amman Madan, Department of Humanities and Social Sciences, Indian Institute of Technology, Kanpur; Nasser Malit, Binghamton University; Dustin M. Wax, Community College of Southern Nevada; and most especially to Dr. Richard Warms, Professor of Anthropology, Texas State University–San Marcos. Their support and the suggestions they offered enriched the book's content and are much appreciated.

Thanks too for their assistance to Dr. Louise Leakey; Dr. Rick Potts, Director, Human Origins Program, Smithsonian Institution; and Dr. Anna K. Behrensmeyer, National Museum of Natural History, Smithsonian Institution; and the staff at the National Museums of Kenya, Departments of Ethnography and Palaeontology and regional museums at Olorgesailie and Sibiloi.

Special personal thanks to Catherine Stover, Johns Hopkins University, Montgomery County Campus library, who located materials and more materials; Norah Olaly, graduate student, Johns Hopkins University, for her help with Swahili; Sally Slater and Dr. Peggy King-Sears, colleagues and personal friends who cheerfully encouraged me; Steve Turner and Peter Liech Adede, Origins Safaris, for providing a wonderful tour of Kenyan sights and sites; Brian and Pat Riley, for explaining the British educational system; Nadine Zangueneh, whose skill with numbers made up for my deficit in math; Steve Vetrano, my editor, who understood when problems arose; the staff at Apex Publishing, LLC, who worked with the manuscript; my daughter Hope McGonigle, for encouraging me to travel to Kenya and for going along; and my husband, Carl Kruhm, for his insights, observations, and encouragement and especially for treating me to dinner when I was tired.

# INTRODUCTION

How and why did early humans come to be?

For many years most people sought the answer through religion. Then in the mid-nineteenth century scientific inquiry led to a flurry of theories and the hunt was on to find the common ancestor that gave birth to modern man and woman.

The abundant fossil record now explains a great deal about human origins, but the interpretation of the fossils and their relationships to each other and to humans today are still unclear. In *Bones of Contention* (1997), Roger Lewin included this truism:

We do not see things the way they are,
We see them the way we are.

He wrote that paleoanthropologist David Pilbeam first thought the quote "came from the Talmud, only later to discover that instead it was from a Chinese fortune cookie." Lewin went on to note, "its source does not, however, diminish its force" (p. 44).

Lewin is correct: People view things from their personal perspective. Each fossil find has been viewed through the seeker's lens, tinted by the desire to see what he or she wants to see and expects to see. And why not? Enormous rewards in terms of fame, money, and power await the finder of the fossil judged by both the scientific world and the public to be the earliest ancestor of modern humans. Obviously, speculation about human origins is in plentiful supply and continues to be fueled by the three C's of Controversy, Competition, and Contention.

Members of the Leakey family, from patriarch Louis and wife Mary, to son Richard and wife Meave, to their daughter Louise, have not wavered in their determination to discover everything possible about human origins. Nor did they shrink from the three C's. This book will describe how the Leakeys, the most famous family in paleoanthropology, have confronted Controversy, Competition, and Contention in their quest to learn how life on planet earth evolved.

# TIMELINE

## Significant Events in the Lives of the Leakeys

| 1903 | Louis Seymour Bazett Leakey born August 7 |
| 1913 | Mary Douglas Nicol born February 6 |
| 1926 | Louis graduates from Cambridge |
| 1928 | Louis marries Henrietta (Frida) Avern |
| 1930 | Louis receives doctoral degree from Cambridge |
| 1931 | Priscilla, daughter of Louis and Frida, born April 13 |
| 1933 | Louis and Mary meet; Colin, son of Louis and Frida, born December 13 |
| 1935 | Mary's first trip to Kenya |
| 1936 | Louis and Frida divorce; Louis and Mary marry December 24 |
| 1940 | Jonathan Harry Erskine Leakey born November 4 |
| 1944 | Richard Erskine Frere Leakey born December 19 |
| 1948 | Mary finds Proconsul skull October 2 |
| 1949 | Philip Leakey born June 21 |
| 1959 | Mary finds Zinj July 17 |
| 1966 | Richard marries Margaret Cropper |
| 1969 | Anna, daughter of Richard and Margaret, born; Richard and Margaret separate and divorce |
| 1971 | Richard marries Meave Gillian Epps |
| 1972 | Louise, daughter of Richard and Meave, born; Louis dies October 1 |

| | |
|---|---|
| 1974 | Samira, daughter of Richard and Meave, born |
| 1978 | Laetoli hominid footprints discovered |
| 1979 | Richard has transplant surgery with kidney from brother Philip November 29 |
| 1983 | Mary retires from fieldwork |
| 1989–1994 | Richard directs Kenya Wildlife Service |
| 1993 | Richard injured in plane crash June 2 |
| 1994 | Richard, with others, forms Safina political party |
| 1996 | Mary dies December 9 |
| 1997 | Richard elected to Parliament |
| 1998 | Richard returns to Kenya Wildlife Service |
| 1999–2001 | Richard serves as senior Kenyan government official |
| 2001 | Meave and Louise announce find of *Kenyanthropus platyops* |
| 2003 | Louise marries Emmanuel de Merode |
| 2004 | Seiyia, daughter of Louise and Emmanuel, born |
| 2006 | Alexia Meave de Merode, second daughter of Louise and Emmanual, born September 3 |

## CHAPTER 1

# A FACT-FINDING MISSION
# (FORWARD TO THE PAST)

> Indiana Jones writes the word *FACT* on the college classroom
> blackboard. He tells the students that archaeology is "...the
> search for fact. Not truth. If it's truth you're interested in, Doctor
> Tyree's Philosophy class is right down the hall."
> —From *Indiana Jones and the Last Crusade*; Boam, 1989

## THE SEARCH FOR FACT

How and when did humans evolve? In paleoanthropology, a first cousin of archaeology that asks this question, facts are found in fossils. Since the late nineteenth century, when prehistory was a new field of study, scientists have slowly collected an impressive array of fossilized bones dating from millions of years ago. They have steadily teased facts that tell a great deal about the overall shape of human origins from these fossils. Many controversies about human evolution still exist but these are mostly in the details of the precise relationships of the fossils to today's humans and to each other.

What facts about human origins do these fossils reveal? Dr. Richard Warms, anthropology professor at Texas State University—San Marcos and coauthor of *Anthropological Theory: An Introductory History* (McGee & Warms, 1996), explains, "All the evidence points to human ancestor divergence from ape ancestors between six [to] eight million years ago. There is much dispute about how and when, but that's the fine details. All the evidence points to Africa, particularly somewhere in East Africa.

Is it possible that this is wrong? Sure. As new evidence comes in, our knowledge gets better . . . but the same could be said about the basic theories of physics" (personal communication, October 25, 2004).

Many paleoanthropologists are hard at work digging for new evidence. There are many ways to dig a hole and to dig deep enough to amass answers about human evolution; paleoanthropologists today look for help from specialists in disparate fields. Molecular biologists, computer technologists, anatomists, linguists, physiologists, and geologists all carry different shovels from their fields to help dig out clues to human origins.

One of today's diggers is Dr. Louise Leakey.

Louise is in her thirties, determined, organized, competent, confident, and also restless because demands, whether imposed by self or outside forces, keep her "incredibly busy" and "run off my feet" (personal communication, March 10, 2004). Although paleoanthropologists work for months in the harsh climate of remote parts of the world, their exploration into human origins is one of the most fascinating areas of anthropology, and the Leakey family epitomizes the romance and allure associated with bones turned into stone.

Louise is the third generation of Leakeys working to flesh out the theory of evolution. She is heir apparent to the dynasty begun in the mid-1900s by her grandfather and grandmother, Louis and Mary Leakey, and carried on from the late 1960s by her father and mother, Richard and Meave Leakey. Today Richard is primarily interested in challenging the world to protect itself and endangered wildlife, but Meave teams in the search for fossils with daughter Louise. In an October 2003 speech at the Louis Leakey Centennial Celebration, Louise called working with her mother "great fun" and said she is "determined we will take this forward."

The story of the Leakey family, however, involves more than simply exhuming dry bones. This is a family whose members, in spite of personal and professional intrigues, indiscretions, and arguments publicly played out, have hungered to find remnants from a world long ago and craved to build a better world far into the future. Their story begins in East Africa in the late 1800s.

## RIDING THE RAIL TO INLAND EAST AFRICA

Foreigners walked the streets and docks of flourishing East African coastal towns since before the common era (BCE), but the few Europeans who traveled inland were primarily adventurers seeking to explore or exploit. As the nineteenth century turned into the twentieth, industrialized countries scrambled to extend their influence and power through colonizing weaker areas of the world, including Africa. In 1887 the Sultan of Zanzibar granted the governing of coastal areas to the British East Africa Association, a private company, but in 1895 the job was handed over to the British crown. Since the group had lost huge amounts of money attempting to control the region, this was not such a generous gift as it might seem. The government renamed it the British East Africa Protectorate; the area officially encompassed what is present-day Kenya but informally included the surrounding areas of Uganda, Zanzibar, and Tanganyika.

British colonial control of the region was gradually achieved by sending troops into the area and by pitting African ethnic groups against each other. The Brits were further aided by smallpox and rinderpest, or cattle plague. Eventually the Africans were subdued and certain areas were designated in which each ethnic group could live, under a British administrator.

The British were determined to unite this colonial outpost with its multiple tribes and ethnic groups. They insisted the people, even the colonial administrators, speak the Swahili language as a *lingua franca* and they built a railroad to link the coastal town of Mombasa to Lake Victoria, a distance of 581 miles diagonally across the heart of Kenya. The single set of tracks was begun in 1896 and completed in 1901, with the work done mostly by 32,000 workers from the Gujarat and Punjab regions of India, another British colony. Its expense to British taxpayers of what would today be almost $380,000 per mile gave weight to its being called a "lunatic line to nowhere" (Pike, 2000, p. 44) by the railroad's opponents in the British Parliament. The way to recoup some of the money, the government reasoned, was to encourage British subjects to move to Africa, with settlement along the railroad.

In the early 1900s the capital city of Nairobi was a village of non-descript small houses and offices clustered along the railway tracks. Soon missionaries began travels to unknown areas of Africa. A. L. Rowse, in *The Story of Britain*, referred to "the astonishing achievements of exploring and missionary enterprise . . ." (p. 146), an accurate description no matter one's point of view. Within a few years merchants and settlers realized that the inland, not the ancient coastal city of Mombasa, was where the future of British East Africa lay. With support from the British government, taxes were raised so white settlers were able to purchase land that indigenous peoples were forced to sell.

Early British settlers were eager for adventure and an opportunity to throw off the suffocating social constraints imposed on them in Edwardian England. They willingly left a society where refined women were expected to be inactive mentally and elaborate outfits forced them to be inactive physically, where well-bred men wore formal attire even walking on the beach during a holiday, and where those who worked for a living were considered servants, not employees. Freed from the rigidity of British society, newcomers to Kenya soon earned a reputation for being slightly eccentric, at least by the conventions of the time. Many of the settlers dressed so informally that one governor refused to eat with constituents who wore pajamas and dressing gowns at dinner.

These few British men and women who braved farming or ranching in inner Africa in the early 1900s were a hardy lot facing huge difficulties. Using farming practices successful at home resulted in crops ruined by diseases unknown in the Western world. The sun scorched plants one day and rain flooded them the next. Horse sickness, tsetse fly, and tick diseases indiscriminately wiped out livestock, including whole herds of cattle and sheep.

Along with adventurers, evangelists flocked to Africa. These missionaries were eager to preach, teach, and heal. Earlier, in the nineteenth century, they had been as unyielding in their efforts to halt trading in slaves as they now were to spread Christianity. From the late nineteenth through the twentieth century, mainstream Western society generally considered proselytizing to be humanitarian, and many Christians, including Louis Leakey's parents, Harry and Mary, answered their church's call to do missionary work.

## THE LEAKEYS

When Mary Bazett married Harry Leakey in 1899, they planned a future as missionaries in Africa. Mary, also known as May, well understood the magnetism of missionary life. Before marriage and against their father's wishes, Mary and her two sisters had responded to the request of the Anglican Church Missionary Society (CMS) to evangelize Africa. The trio traveled to Mombasa, along the southern Kenyan coast. Ill health forced Mary's return to England, where doctors advised her never to return to the tropical climate of Africa.

Harry and Mary Leakey decided to disregard the medical advice given her and so, in December 1901, Harry left his position as curate in a London parish to settle at a small mission in Africa and prepare for his family's arrival. In the spring of 1902, Mary once again sailed to East Africa, this time joined by her two young daughters, seven-month-old Gladys and three-year-old Julia, and Miss Oates, a caregiver for the children.

Rather than coastal Kenya, where Mary's health suffered on her previous trip, the family lived inland in hilly central Kenya, at a mission in Kabete. The peak of Mt. Kenya rose to the northeast, Mt. Kilimanjaro was to the southeast, and the new and nearby town of Nairobi lay halfway between these peaks, slightly to the southeast.

After a trip of almost a month by steamship and then train, the last few miles to her new home at the mission must have seemed jolting both emotionally and physically to Mary. Harry, anxious to see his family, met them in Nairobi and traveled on horseback back to Kabete, but the rest of the family was slung in hammocks with each end carried by Kikuyu men who wore little and sang much as they made their way along meandering rain-soaked footpaths to their new home.

The Kabete mission was located in the midst of nature at its finest. Situated on approximately sixteen acres of land purchased from indigenous families, the mission provided sanctuary to Africans who were ostracized by their families when they converted to Christianity.

The Leakeys were among the first missionaries to travel to the interior of East Africa and moved there with a desire to do more than convert. They evangelized while also establishing schools and providing access to Western medicine with great but gratefully given personal sacrifice. Mary started a school for girls and was later recognized for her work in education when a school was named after her. Both Leakeys won the respect and admiration of the local people by their desire to serve and the obvious love Mary and Harry had for Africa. Harry's nickname of *Giteru* referred to his having a big beard. Mary was called *Bibi* (Beebee), a term that indicated respect and honor, sort of the female version of *Bwana*, the term for a European male.

Although they returned to England for long visits of a year and sometimes more, Louis Leakey's parents spent most of their lives in East Africa. Harry died in 1940 and Mary in 1948.

## LOUIS: AT THE FRONT OF THE QUEUE

The third child of Harry and Mary, Louis first opened his eyes on August 7, 1903, in a home that was a great distance from England in both miles and amenities. His birth was premature and in *White African*, one of several autobiographies he wrote, Louis called his survival "nothing short of a miracle" (L.S.B. Leakey, 1966b, p. 4). The house was made of mud, with a dirt floor that turned into mud when rain leaked through the poorly constructed thatched roof. The church mission, although located near the equator, was 6,500 feet above sea level and was cold and drafty when the sun set. With no glass available for windows, shutters helped keep out cold night air and a smoky brazier provided some warmth.

A decent small stone house wasn't available to the family until 1905 when they returned from England on a trip mandated by Harry's poor health. Douglas, the youngest child of Harry and Mary, was born in their new Kabete home in 1907.

During his early years Louis and his sisters were bilingual and easily interacted with the Kikuyu. Louis roamed freely in the woods near the mission, with its abundance of birds and animals, and played games with toy bows and arrows. The children had a tutor in the morning and after tea took a nature walk.

The typical sequence for Church Missionary Society families was to spend three years in the field and then, since the trip was long and difficult, go back to England for a year. The Leakeys returned to England in 1911 for this one-year furlough but, because of Mary's health problems, spent longer than an additional year there. Harry Leakey, who had once been a schoolmaster, used the time to start a preparatory school. Louis and three other boys were the first students. This interlude in England provided the framework of a British education that was continued by Miss B. A. Bull, a classically trained tutor who returned to Africa with the family in May 1913.

Miss Bull expected the Leakey children to focus on learning Latin, but they were more interested in watching a life-or-death fight between a chameleon and a tree snake. Study time often evaporated. After tea Miss Bull accompanied the children while they climbed through huge old hollow tree trunks and across a natural bridge to visit imaginary people with whom they romped until it was time to collect wildflowers for their mother and return home. Miss Bull's love of the classics could not stand up to the Leakey children. They choose escapades over education and she soon quit. Her replacement, a Miss Broome, endeared herself by creating fun school activities.

## THE GREAT WAR

In 1914, World War I began and spread around the globe. Great Britain, France, Belgium, Russia, Italy, Japan, and in 1917, the United States, joined in the fight against the imperialism of the countries of Germany, Austria-Hungary, Turkey, and Bulgaria, known as the Central Powers.

Louis was only eleven years old at the start of the war. Like many white Kenyans, he had planned to continue his formal education in England when he was thirteen, but World War I both kept him in Africa during his adolescent years and affected his daily life.

Because countries on both sides of the conflict had colonies on the African continent, rumors roared through the countryside. A contingent of all men able to serve volunteered to defend the British East African border from any troops attempting to cross from German East Africa (now Tanzania). Missionaries organized the Volunteer Carrier Corps, and hundreds of young Kikuyu men were soon writing home to request snuff, their favorite form of tobacco, and books to continue their education and prepare them for baptism. Louis and others at the mission busily deciphered letters for family members who could not read and write and addressed packages and letters to be sent to the war zone.

Although the Volunteer Carrier Corps was not involved on the front lines, local casualties were numerous and Louis and others at the mission were responsible for giving the sad news to relatives. Many of the casualties were a result of illness. As horrific as the war itself was, the movement of soldiers and ships set in motion a global epidemic, or pandemic, with at least twenty-one million people around the world dying from the Spanish influenza. Whole Kikuyu villages were wiped out and men returned home from the battlefields to sadly find their entire family gone.

## TEEN YEARS

Louis recalled the years from ages ten to sixteen as filled with happy incidents. Because he wanted the same early independence from his family that his Kikuyu friends enjoyed, Louis successfully built a hut for himself at age eleven. It had one room about eight feet by ten feet, rather like a ground-bound tree house. He and his siblings had much fun and even cooked potatoes from their own vegetable gardens. At age thirteen and a man by Kikuyu custom, Louis hired two African laborers and built a second house. He purchased supplies with money earned by selling the skins of wild animals and also selling zoos live birds and small animals that he trapped. Louis's parents didn't consider this second house safe enough for him to live in, so a year later, in the Kikuyu tradition of helping friends complete homes after basic construction is done, he enlisted friends to help him build a third house, where he lived except, at his mother's insistence, for dining with his family.

Mary tried to enforce the Western tradition of parents knowing where their children are, but Louis craved the independence of his

Kikuyu friends and spent his days helping them build their houses, observing and trapping game, caring for pets, and playing British football (soccer, to Americans) and Kikuyu games. One tribal game involved bowling a hoop while team members threw their spears at the moving hoop and included "prisoners" captured from the opposing team. The rules were complicated and Leakey noted in *White African* that the game was valuable to the Kikuyu understanding of tribal warfare, customs, and social behavior (L.S.B. Leakey, 1966b, p. 28).

Animals, both wild and family pets, played a large role in Louis's life. He became friends with Joshua Muhia, a Kikuyu man who was a hunter and trapper rather than a farmer, like most Kikuyu. Muhia taught him how to make cord by peeling bark from plants, separating hard outer from soft inner bark. This inner bark was chewed until it was frayed and then rubbed against the thigh until it was a satisfactory cord with which to make a spring noose trap. Muhia also taught him how to catch larger animals, like hares and duiker, a small African antelope.

While on these hunting expeditions, Louis picked up bits of obsidian, formed by molten lava and similar chemically to granite, and recorded information about it. His interest in stone tools had begun Christmas 1915, when his cousin in England sent a book that influenced his life. *Days Before History*, by N. H. Hall, described European Stone Age tools, important because they show the mind at work and are the first lasting evidence of complex behavior from early humans. Although those described were made of flint, Louis immediately found tools with a similar appearance around Kabete but received little support for his finds. His parents focused on the type of material rather than the shape and, realizing the tools were not made of flint, questioned their value. His Kikuyu friends argued that they dropped from the sky after a rain, a logical if incorrect conclusion, since hard rains washed them from the soil where they had hidden for millennia.

No one except Louis had much enthusiasm for the tools until he showed them to Arthur Loveridge, a family friend who later became a professor of herpetology at Harvard University. Loveridge assured Louis the rocks he had found were in fact ancient tools and advised him to keep a record of his finds. Flint, unavailable in Europe, doesn't exist in Africa and so Stone Age peoples there made tools from obsidian, which actually provides a sharper edge and is easier to work with than flint, although it breaks more readily. The African version of these hand axes went unrecognized for many years by Western archaeologists, who, like Louis's parents, rejected them because they weren't the dull grayish-black of flint. (See chapter 4 for more on Stone Age tools.)

Loveridge also encouraged Louis to turn his interest in trapping animals and birds to categorizing and studying them. Louis considered a career as an ornithologist but, while his interest in birds was life-long, his first love was prehistory. These early teen years moved Louis toward a career as an anthropologist, gave him skills and experiences that served him well as an adult, and resulted, during his college years, in Louis's decision not to follow his parents into a career as a missionary.

With the global spread of World War I, Miss Broome left to nurse battlefield casualties and so Louis's formal education was unstructured until he returned to England when the war ended. By this time, the gap between him and English teens his age was socially and academically huge, but for the rest of his life he used the practical skills learned in Africa.

## BETWEEN TWO WORLDS

A 1919 armistice brought peace to the world and, at age sixteen, Louis left Africa for school in England. He looked like a Westerner, had spent some of his childhood years in England, and had the broadened perspective born of a cross-cultural upbringing, but in January 1920 when Louis entered Weymouth College to prepare for entrance to a university, he quickly discovered he didn't have the academic background or the culture young Englishmen had shared since birth. He had never been to the theater and had no idea how to write an essay. Knowing the cultural niceties of negotiating with the Kikuyu for a tree to use for beehives, the etiquette involved in sharing the hives with those who helped, and the "ndumbi" and "mbogoro" methods of hanging the hives was not useful information at Weymouth. He played sports in Kenya but neither understood the game of rugby nor liked wearing the regulation heavy shoes. He later described himself as being the butt of verbal abuse from both teams when he wandered around the field and tried to understand the rules. Louis made it obvious to his classmates that he felt they were childish and this attitude added to his being ridiculed, bullied, and seen as arrogant and different.

In 1920 he fretted about being unpopular and feeling ill-at-ease among his classmates, but by the time he penned *White African* in his early thirties, he wrote that he was surprised he fit in as well as he did (L.S.B. Leakey, 1966b, p. 72). In England, rules such as when to eat, when to study, and when to go to bed controlled every minute of his life. These rules chafed Louis and seemed ridiculous. After all, in Kenya he had built his own house in which he lived and he had a tremendous amount of freedom to spend time outdoors, hunting and trapping as he wished.

Through the rest of his life Louis moved in several different worlds. His first autobiography is aptly titled *White African* and, although generally considered a British archaeologist, Louis identified himself with Africa rather than Great Britain, perhaps because he felt like an outsider in England and claiming East Africa as his homeland gave him a unique identity. Louis began *White African* with a quote by Chief Koinange stating that Africans consider Leakey one of them rather than European and refer to him as "the blackman with a white face" (L.S.B. Leakey, 1966b, p. 1). This description flattered Louis and he "always considered myself more of a Kikuyu than an Englishman in many ways" (p. 1) and "anything but typically English" (p. 71). In his autobiography, *One Life*, son Richard stated Louis was an African, but a "little formal and westernized" (1986, p. 19) and "did not fit easily in English society" (p. 16). Carolyn Clark, who studied Louis's relationship with the Kikuyu, suggested that Louis could show several different persona at the same time, and that, like African masked dancers who "embody the spirit of the mask they don Leakey could mix Kikuyu conservatism, British nonconformity, colonial outrage, or Kikuyu opportunism, British loyalty, and colonial paternalism" (1989, p. 383). Science writer Roger Lewin wrote simply that Louis was "a maverick by any standard" (2002, p. 84).

## CAMBRIDGE, ST. JOHN'S COLLEGE

When he entered Weymouth, Louis anticipated eventually being accepted as a student at Cambridge University even though he would need a scholarship to pay his tuition. Hoping for help, he tried to discuss his goals with the headmaster but received only a shrug and the comment that he should try for a position in a bank. As he wrote years later, Louis was "furious" and "utterly miserable" (L.S.B. Leakey, 1966b, p. 75) and had no intention of giving up his goal of receiving a degree from Cambridge.

British organization of higher education is different from that in the United States. US universities are made up of various colleges that provide instruction within the same field; the College of Engineering, for example, offers courses only in engineering and the student goes to the College of Arts and Sciences for courses in history, biology, and English. At Cambridge, colleges such as Peterhouse, Corpus Christi, and St. John's are self-contained institutions, but all students receive a degree from Cambridge University regardless of the college they attend. Louis's father had attended Peterhouse College and other members of his family attended Corpus Christi. Mr. E. A. Benians at St. John's College encouraged him to apply there and also advised him about obtaining a grant.

Stiff entrance examinations in mathematics and three languages were required and Louis passed Latin and French in addition to math. He also obtained letters certifying his knowledge of Kikuyu, since an examination in Kikuyu wasn't possible. The letters were from the head of the Church of Scotland Mission and a Kikuyu chief, probably, Louis reflected, some of the most "curious certificates" (L.S.B. Leakey, 1966b, p. 77) ever offered. Louis began his career at Cambridge in October 1922, armed with a £40 grant, barely enough to pay expenses for one year if he were frugal. Having just enough money to scrape by became a pattern for the rest of Louis's life.

## FROM UNFORTUNATE ACCIDENT TO POSITIVE INCIDENT

Louis's first year at Cambridge felt freeing after the restrictions at Weymouth. He failed the final examination in French literature, a blow to his hopes for a scholarship, but Louis was always creative in cobbling together needed funds. In this case, money from a summer spent cooking and housecleaning for a cousin, a gift from his parents, continuation of his previous grant of £40, and a second grant available to children of missionaries who themselves planned a missionary career allowed him to return to Cambridge. During his last two years he bought over a hundred ebony walking sticks from Kenyan woodcarvers and traded them for clothes in England.

Louis wanted to win his college colors in rugby during his second year at Cambridge. Then, at an important match in which he was playing at the height of his game, Louis was kicked in the head. Given the nature of Louis and the nature of rugby, he insisted on returning to the game and was quickly injured again. The next day any attempt to focus resulted in a terrible headache. Doctors finally ruled that he had to give up college for a year. This prescription was a terrible blow, since clearly Louis could not afford a yearlong holiday. Dizzying headaches forever plagued him when he overworked, but Louis turned the accident into an incident that was the driving force the rest of his life.

# CHAPTER 2

# IN THE BEGINNING . . .

## *(Louis, 1925–1933)*

> The past is the key to our future.
> —Richard Leakey, quoting his father,
> Louis, in the prologue to *Origins Reconsidered*, p. xv

## RETURN TO EAST AFRICA

Louis still planned to follow in his parents' missionary footsteps but he needed money to survive while he recuperated from his rugby accident. The job he found while still a Cambridge undergraduate marked the beginning of a long career in anthropology. A family friend told him that the British Museum of Natural History was sending an expedition to Tendaguru in Tanganyika Territory (now Tanzania) to search for a dinosaur skeleton. Its leader, W.E. Cutler, was known for collecting dinosaur specimens in Canada but, not knowing the language or the landscape of East Africa, he needed someone with expertise and experience in both.

Louis was hired and on March 17, 1924, after the welcome experience of traveling first class for the first time, he and Cutler landed in Mombasa and began the trek to Tanganyika to build and equip a camp that could accommodate a hundred workers. Louis busily dealt with government officials, organized porters to carry supplies, supervised the construction of buildings, negotiated with the locals for some fresh food, and secured and rationed water and additional food and supplies. During the six months he was with the expedition, Louis learned from his boss how to get fossils safely out of the field, especially Cutler's well-honed technique of coating them with plaster. Cutler tried to surround the process with an air of mystery and insisted on doing all the plastering himself, and Louis, forced to stand and watch, surely must have felt frustrated.

Cutler must have been frustrated himself, since he was in his forties and had to tolerate a know-it-all half his age who, in fact, did know it all, relative to Cutler's skills with Kenyan culture and language. Cutler was dependent on Louis both to direct and protect him and to supervise the workers in their language. Soon after arriving in East Africa, Louis warned Cutler to avoid a plant whose pods had tiny hairs that were painful to human skin. Cutler ignored the warning, wrapped some pods in a handkerchief, and later forgot the warning and used the handkerchief, filled with irritating hairs from the pods, to wipe his face and neck. Louis supplied lotion that soothed Cutler when he was "dancing about in agony, yelling like a madman and cursing like a trooper" (L.S.B. Leakey, 1966b, p. 101).

Along with a variety of ailments and tropical diseases, including malaria and dysentery, both Louis and Cutler suffered each other's idiosyncrasies and, amazingly, when the six months of his contract ended, Cutler wanted Louis to remain with him. Cambridge, however, expected him back and Louis returned in January 1925.

Even though Cutler worried Louis would steal his glory by publicizing their adventures before Cutler himself could make his way back to England, Louis received permission to earn extra money by writing articles and lecturing.[1] In *White African* he wrote, "As I was only an undergraduate in my second year, and not yet twenty-two years old, and never having lectured in public before, I grew more and more worried as the day approached" (p. 129). He was rightly terrified. Not only did he have to lecture before a large crowd that included Cambridge and British Museum dignitaries, he also had to wear formal attire, including tails, similar to that worn today by grooms at weddings. Given Louis's finances, he borrowed clothes and they didn't quite fit.

As Louis spoke, he relaxed, his first lecture ended a success, and he

soon began to earn extra money by public speaking. He had an unlimited supply of stories to tell school students, like about the shadowy
animal he assumed was a leopard that eased into his bedroom and then
grabbed his pet baby baboon and escaped with it through the window.
And in describing walking in the bush without his gun and being surprised by an angry cow elephant protective of her young calves; fearing
that the elephant would sense his closeness and charge him, Louis later
wrote that "with thumping heart I very slowly crawled away" (L.S.B.
Leakey, 1966b, p. 107).

Although he always suffered a few minutes before beginning a lecture, Louis discovered he enjoyed public speaking and his lecturing
earned a tidy sum that helped cover college expenses and later was crucial to fund-raising efforts for fieldwork.

Louis always referred warmly to his adventures with Cutler, and
they influenced his future in important ways. He gained tremendous
experience in organizing and directing an expedition and an understanding of how to excavate fossils from the field. He now became certain he wanted anthropology as a vocation, not an avocation.

First, however, he had to pass his examinations in French literature
and Kikuyu. On his return to Cambridge, Louis spent most of his time
studying French and passed. He particularly enjoyed telling the story
about how he tested himself in Kikuyu. Before the university agreed to
his using Kikuyu along with French to satisfy his language requirement,
they checked that two Kikuyu speakers were available to test him but
did not require their source to give the two names. Louis was amazed
when the final examination questions sounded familiar. Eventually he
discovered that one speaker referred to Cambridge was named Louis
Leakey. When Cambridge found their own student was one of the two
Kikuyu experts in England, they obviously needed another Kikuyu
speaker, so they tapped Louis's "tutor." The tutor didn't speak the language and in actuality learned it while Louis ostensibly was working
with the tutor to maintain his fluency. The story spread that Louis
examined himself in Kikuyu.

Given his brashness and background, Louis had several run-ins
with Cambridge authorities. In one instance, when doubters questioned
the power of the drum for communication between African villages separated by several miles, Louis stationed friends at various distances

around the university and in the town, climbed on the dormitory roof with his drum, and boomed out a signal. The walls of Cambridge reverberated with the deafening sound! Louis quickly moved the drum to his room, placed coffee cups on top to disguise it as a table, and was quietly reading by the time authorities checked.

The powers that be also ejected Louis from the tennis court and reprimanded him for daring to wear shorts and a loose, African-style "bush" shirt they deemed indecent. Long white flannel trousers and white shirts were *de rigueur* on Cambridge courts during the 1920s. A photograph of him with about three inches of knee showing makes it easy to appreciate his feeling that the indecency claim was absurd, especially since rugby players in uniform revealed far more.

## THE MOVE TO ANTHROPOLOGY

Now focused on a career in prehistory, Louis was fortunate to spend his last two years as an undergraduate studying with the eminent anthropologist, Dr. Alfred Cort Haddon, who had inaugurated coursework at Cambridge in the new field of anthropology in 1904. Haddon was considered a brilliant teacher and he also went beyond classroom instruction and invited students to his home on Sunday afternoons, where they had access to his extensive library and enjoyed informal discussions. Another student who benefited from Haddon's hospitality was Gregory Bateson, who later became well known both as an anthropologist and as the third husband of Margaret Mead, a famous American woman who popularized anthropology in the United States through her writing and lecturing from the 1930s through the 1970s.

During these Sunday afternoons Louis was introduced to what became a lifelong interest in string figures. The children's game of Cat's Cradle represents a simplistic version, but many of the creations then as now were complex and illustrated stories among many different ethnic groups. String games were not popular among the Kikuyu, so Louis was unacquainted with them but soon found they could help win over almost any audience, child or adult, in any culture, and he was known to carry a length of string in a pocket to use as an icebreaker.

As Louis's undergraduate days at Cambridge ended, he made plans to carry out excavations in Kenya. Most of his professors at Cambridge expressed doubt and one was completely pessimistic that the evolution of humans began in Africa.

## EVOLUTION HAPPENS

At a time most Christians believed literally that the world was created in seven days, the entire first edition of Charles Darwin's *The Origin of Species* sold out on the day it was published, November 24, 1859. The book, controversial and important in that order then, remains important and controversial, in that order, today. Although religious leaders initially had great difficulty with the concept of evolution, change of opinion between both clerics and laypeople of mainstream religions gradually evolved when they realized Darwin's basic ideas allowed belief in a creator's initiating the entire process.

Actually, evolution was more an idea whose time had come than it was a brand-new concept originating from Darwin. In an era of scientific explosion, Darwin himself counted twenty others who preceded him in writing about evolution, including his own grandfather. Darwin's name is most closely linked to the theory because he studied, reflected, and wrote about the topic for thirty years and added natural selection to his meticulous observation and documentation of evolution. The respected scientist Thomas Henry Huxley, who became a zealous supporter of Darwin, probably echoed the thoughts of many scientists when he commented that he was stupid not to have thought of it himself!

Evolution is not a theory in the sense that it is open to speculation. Change happens. Evolution happens. Evolution, or change in a species very slowly over time, is not preplanned. Events happen and some individuals die as a result of them and some do not. Those that survive pass on their genes to their offspring. Sometimes new genes arise by chance and get passed on. Each generation passes on something a little different, but change is not random. Over many, many, many generations, large changes are evident.

Evolution is itself undeniable although the details are still being worked out. Anthropologists can say with certainty only that humans have, over several million years, evolved. They now believe human divergence from apes happened in Africa, probably eastern Africa, about 6 million years ago, give or take 1.5 million years, and these early humans moved throughout the world 1 to 2 million years ago. Could this specific scenario be wrong? Of course. Science is a matter of continual investigation, with old assumptions replaced by new as more facts are learned, but right now that is what the fossil record, supported by research from other fields, shows.

## FINDING THE ORIGINS OF MAN

The question of where humans originated has been called the "ultimate conundrum" (Lott & Smith, 2002, p. 29). Charles Darwin himself suggested Africa as the cradle of human civilization because Africa is the home of apes and chimps, the closest relatives to humans. Scientific wisdom in the 1930s, however, pointed elsewhere. Few Westerners were willing to be linked to apes, and, while scientists accepted evolution, few acknowledged the possibility that humans had their genesis on what was widely called "The Dark Continent." More acceptable was the theory that people evolved separately from each other in different parts of the world. Because the races looked so different, this assumption seemed logical, and Leakey stood unwavering and almost alone. He returned to Africa because his first stone tools, found at age thirteen, convinced him he would find fossils that proved modern humans split from the great ape family and then spread from Africa throughout the world.

One other scientist who believed human prehistory began in Africa was Raymond Dart. In 1924 Dart discovered the skull he named *Australopithecus africanus*, or Southern Ape of Africa, but it was called Taung Baby because of the site in South Africa where it was found and because it was little bigger than a man's fist. Actually, the skull was sent to him in a box with interesting hunks of limestone but, with experience in anatomy, Dart recognized its value. Few others in the scientific world did. During the early half of the twentieth century scientists believed humans had recently evolved in Europe or Asia, so geographically Taung Baby had a strike against it. They also believed brain size evolved before humans walked upright and this little fellow had a very little brain. Besides, a child didn't seem like a good specimen on which to hang a species. Dart was ignored, criticized, and made the butt of jokes by scientists and the public in general for several decades because the skull simply did not fit into the preconceived view of human evolution. Louis didn't seem impressed by Taung Baby but he never believed australopithecines were human ancestors anyway. He later wrote, however, that he always thought Dart would be vindicated (L.S.B. Leakey, 1974, p. 22).

## RETURN TO KENYA

In July 1926 Louis and Cambridge friend Bernard Newsam sailed for Kenya to begin "The East African Archaeological Expedition," a "high sounding title," Louis later acknowledged (1966b, p. 156). He planned to carry out a brief investigation of several sites, including the Gibberish cave where he played as a child. When he was small he hoped to find Stone Age tools lying on the floor of the cave. Now he realized they would need to be excavated, since the soil in which they lay was not exposed to erosion by the elements. By the expedition's end he had identified more than sixty sites for investigation, many of which are still being searched. Louis and Newsam first set up camp at Nakuru, about one hundred miles west of Kabete. Then when Louis spied a skull projecting from the soil at Elmenteita, the team immediately moved their camp there. Quarters were, Louis wrote, "very large and airy" (1966b, p. 165), a phrase that puts a positive spin on being housed in an old pigpen. Although living in a pigsty doesn't sound luxurious, this camp offered conveniences. Louis had recently purchased a used Model T Ford and fresh water could be easily carried back to camp by car. The excavation site was near the stream so the team could bathe at the end of the workday.

While at Nakuru, they found fish bones, evidence that the climate had changed considerably in the last two million years, since other possible sources of water were either too laden with minerals or too far from the site. Dr. Erik Nilsson, a Swedish geologist, visited their camp for six weeks and provided Louis and Newsam with valuable information and advice. On January 4, 1927, Louis found a second skull, additional bones, and tools and pottery at Elmenteita; these artifacts led him to believe that the people who lived at Elmenteita were not so sophisticated in their development as those at Nakuru, but from similarities in the artifacts, he sensed the two sites were linked. From the evidence he was collecting, Louis had already begun reaching important conclusions about the evolution of culture in East Africa.

Louis was also invited to explore two caves, or "rock shelters" on the farm of a Mr. Gamble, about ten miles away from Elmenteita, logically dubbed Gamble's Cave by Louis. They found a trove of pottery, stone tools, and bone fragments waiting to be explored, but with funds now low, Louis returned to England and immediately began planning a return visit to uncover "all the secrets hidden in the deposits" (1966b, p. 170).

Louis was certain he had made discoveries that could have important consequences but also worried that he could not raise additional funds if he were misjudging their value. In sum, however, he wrote that he was pleased and felt he had achieved "far more" (1966b, p. 171) on the first full expedition he planned and carried out than he expected.

Louis made an additional discovery while on this expedition. Three visitors stopped by Gamble's Cave. One visitor was Henrietta Wilfrida Avern, or Frida, as she preferred to be called. A Cambridge student, Frida had taken several archaeology courses and was charmed by Louis's attention, while Louis was delighted to discover a bright, excited listener for his enthusiastic descriptions of his work. They corresponded after Frida returned to England and, when Louis proposed, they were married in July 1928. The Averns were a conventional upper-middle-class family and not terribly enthusiastic about Louis's unconventional background, but Frida was enticed by the promise of an adventurous life with Louis. He was planning his Second East African Archaeological Expedition and invited Frida to join him.

## EAST AFRICAN ARCHAEOLOGICAL EXPEDITION #2

Louis appreciated intelligent women and accepted females in fieldwork at a time when most men did not. He was unique in collaborating with

his wife and opening his expeditions to other women. Louis invited
Penelope Jenkin, a friend of Frida's who was looking to use her expertise in the ecology of lakes, to take part in the second East African
expedition. Penelope thought inviting her was a kindness typical of
Louis and realized her involvement required work on his part (Morell,
1995, p. 40).

To Louis, life at the Elmenteita camp was "almost as humdrum as
it might be in a suburb of London" (1966b, p. 189). To stir up excitement for new arrivals, the group staged a lion's roar and a raid by
Maasai warriors.[2] The roar of the lion and the noise of the Maasai were
faked, but the birds and other wildlife were not and were spectacular.
Penelope Jenkin described seeing what she thought was a flight of star-
lings over Lake Elmenteita, only to discover as they flew lower that they
were white pelicans with brilliant yellow bills (Morell, 1995, p. 40).

In eight months at Elmenteita, from September 1928 to April 1929,
Louis's team found two *Homo sapiens* skeletons and a variety of artifacts. Then, to Louis's great delight, just before they closed down the
camp, geologist John Solomon and Elizabeth Kitson excavated a pear-
shaped stone hand axe, probably serving multiple purposes for early
humans, that Louis estimated to be about 40,000 to 50,000 years old.[3]
When Solomon and Kitson found the first hand axe, they weren't sure
what it was. Louis recognized its value immediately and sent the two,
skeptical about the possibility of finding more, back to the deposits the
next day to continue searching, and they did discover additional implements. (See chapter 4 for more on hand axes and similar stone tools.)

Feeling quite pleased, Louis, Frida, Elizabeth Kitson, John
Solomon, and Jannai Kigamba, a Kikuyu worker talented with car
repair, left for the annual meeting of the British Association for the
Advancement of Science in Johannesburg, South Africa. Taking
Kigamba was a wise move. Some sections of road had ruts that could
be measured by feet rather than inches. Louis noted in *White African*
that in 1929, one stretch of road was especially terrible, although by
1935, when he wrote the book, "one could easily proceed at 25 instead
of 5 miles an hour" (1966b, p. 204).

Louis was notorious for dating finds as far in the past as possible
and publicizing his work widely. In *Ancestral Passions*, Virginia Morell
wrote, "His insistence on giving his sites absolute dates and always

pushing for dates of the greatest antiquity, however implausible, became a pattern on these early expeditions, as did his propensity for grandstanding and overstatement" (p. 44). Knowing Louis well, Professor Haddon wrote to caution him not to overestimate dates or try to impress his colleagues at the conference. Louis must have followed Haddon's advice because his presentation about the Elmenteita explorations was so well received that many of the scientists wanted to see his sites for themselves before returning home. Although the trip to Johannesburg took six weeks, Louis and his team rushed back in a short two weeks to prepare for their guests. The sixty visitors were impressed with Louis's work. A reputation that now glowed brightly arrived back in England before he and Frida did.

Louis stayed in Kenya several additional weeks. Concerned as always with the plight of the indigenous Africans, he agreed to work on a governmental committee studying Kikuyu land acquisition. He hoped his membership could help protect Kikuyu lands from takeover by the British, who misunderstood tribal property ownership and Kenyan mind and mores.

By 1930 the couple was back in England. Louis received his doctoral degree and their first child, Priscilla, was born in 1931.

## EAST AFRICAN ARCHAEOLOGICAL EXPEDITION #3

Louis had barely arrived back in England before he began to plan his third expedition, this time to Olduvai, then known as Oldoway, Gorge in Tanganyika, now Tanzania. He had earlier examined a skeleton found there in 1913 by Dr. Hans Reck, a German paleontologist. Its age was a source of controversy among anthropologists and Louis, feeling it was similar to the Elmenteita skull, wanted to visit the Olduvai site. Reck, who assumed after Germany's loss in World War I that he would never return there, was delighted with Louis's invitation to join the expedition. Reck had searched without success for hand axes in 1913 and, believing none would be found, agreed to a friendly wager. Within an hour after their arrival at Olduvai, a Kikuyu worker found a hand axe. Like the European archaeologists and Louis's parents many years before, Reck had ignored shape and searched for tools made of flint on his expedition eighteen years before.

To Louis's disappointment Frida refused to stay at Olduvai with him. No matter how exciting Louis found fossil finding, her baby took priority over fieldwork for Frida, although it was difficult for her to live three hundred miles away from her husband. The Reverend Harry and Mary Leakey built a home in Limuru when they retired and Louis constructed a three-room house on their land for Frida, baby daughter Priscilla, and a friend. It was really little more than a hut with a corrugated metal roof and dirt floor.

The expedition's primary goal was to date Reck's Olduvai skeleton and, to everyone's surprise, Reck easily found the four wooden pegs he had used to mark the site. They also found a trove of ancient animal fossils and hand axes, but no human fossils to corroborate the age of Reck's skeleton, so dating was determined by reading the soil where Reck had excavated eighteen years before. Because Louis saw similarities between Reck's skull and his Elmenteita skull, he believed they dated to the Mesolithic period, or middle Stone Age, and that both were about 10,000 years old. Based on the geology of the bed, Reck convinced a doubting Louis that the skeleton was nearly 500,000 years old and Louis, once convinced, widely and forcefully promoted the skeleton as the oldest *Homo sapiens* ever found. As biographer Sonia Cole pointed out in *Leakey's Luck*, Reck "must be one of the few people who succeeded in swaying Louis once his mind was made up" (p. 88). Which puts a rather positive spin on it since, as Richard Warms, anthropology professor at Texas State University–San Marcos, notes, "Reck succeeded in swaying Louis in the direction of increasing publicity and fame for Louis . . . perhaps not such a difficult job after all" (personal communication, August 3, 2004).

In the early 1930s dating techniques were still in their infancy, but geological tests conducted in England on soil samples and on the skeleton itself found what later carbon-14 tests confirmed: Reck's Olduvai skeleton was only about 17,000 years old. (See chapter 7 for more on dating techniques.) Unfortunately, support of Reck was one of a number of hasty conclusions Louis rushed to announce publicly and in professional journals. Such rapid conclusions made his later pronouncements somewhat suspect to the scientific community. Although Louis eventually accepted the more modern age of Reck's skeleton, he was still convinced that the human story began in Africa.

After some of the other members of the expedition returned to England, Louis continued to search at Olduvai, Kanam, and Kanjera. At Kanjera fragments of three *Homo sapiens* skulls were found along with hand axes. Even better, the fragments were *in situ;* that is, actually peeking from the earth where no one could dispute their location. The fragments seemed to indicate a larger brain than skulls found previously and Louis deduced that the Kanjara fossils represented evolution from a lower jaw, unfortunately not seen *in situ*, found and collected by his chief fossil finder, Juma Gitau, at Kanam and dubbed Kanam Man, or *Homo kanamensis*, by Louis.

## BEING ON GOOD TERMS IS IMPORTANT

The biological terminology requires some explanation. The taxonomy has been modified but is still basically the one developed by Linnaeus in 1758 and based on observable physical characteristics. Scientists in Leakey's era placed all living things in seven categories: kingdom, phylum, class, order, family, genus, and species, with the latter three important to human, or *hominin*, fossils. (An eighth taxonomic level, *tribe*, has recently been introduced because molecular evidence has bumped morphology, but that's another story.) Pretty much because they walk on two legs (that is, are bipedal), humans belong to the family *Hominidae*, then to the genus *Homo*, or human, and finally to the species *sapiens*, meaning *wise*, perhaps debatable. With the name *Homo kanamensis*, Louis planted Kanam Man in the human genus and assigned it to a new species.

Louis wrote, "During the same field season in which we had found the Kanjera skulls, we had found a hominid lower jaw at Kanam West" (L.S.B. Leakey, 1974, p. 20). Today he would write that a *hominin* lower jaw was found. The terms *hominin* and *hominid* have changed over time. Until the 1980s humans were considered different from other apes, so *hominid* referred to humankind's immediate ancestors—the meaning *hominin* has today. The term *hominid* now has a broader meaning. This change in terminology has been gradual and Web sites, books, and anthropologists in person may vary in the use of these terms. For consistency, in this book:

*hominin* refers to modern and extinct humans, and human ancestors, including members of the *Homo*, *Australopithecus*, *Paranthropus*, and *Ardipithecus* families but does not include the
lesser apes; that is, orangutans and gibbons.

*hominid* refers to all modern and extinct humans, chimpanzees,
gorillas, gibbons, and orangutans and their immediate ancestors.

Quotes, however, are not changed and reference may relate to the
older meanings.

## AND NOW, BACK TO OUR STORY

Back in England in December 1932, with applause from his colleagues
for his Kanam-Kanjera fossil finds, Louis felt he was on his way to realizing his life's dream—finding prehistory's roots and the beginning of
human evolution in Africa. He enjoyed a rising career at one of the
world's most prestigious universities, support from the scientific community (despite his error in supporting Reck), the pleasure of being a
father, and a wife who loved him. But his life was about to change.

In April 1933 Frida became pregnant with their second child. It was
a difficult pregnancy. She was bedridden much of the time and, with a
two-year-old to also care for, her work in anthropology took a back-
seat. Louis began to live a life separate from his wife. Rather than
returning home, he often slept in a small suite of rooms Cambridge gave
him to store his stone tools and fossils and he made frequent trips to
London to work for several days at the Royal College of Surgeons. He
increasingly felt dissatisfaction with what he considered a hasty marriage and was involved with several women. Frida, preoccupied with
her daughter and unborn child, didn't realize Louis was less than happy
with their marriage and was oblivious to Louis's comings and goings.

After Louis spoke one evening at the Royal Anthropological Institute, Gertrude Caton-Thompson arranged for a young archaeologist
and artist friend named Mary Nicol to sit next to Louis at the dinner
party that followed. Caton-Thompson thought Mary would be an
excellent choice for illustrating Louis's books and seized on the opportunity to introduce them. The introduction was one Caton-Thompson,

a friend of Frida as well as Mary and Louis, soon regretted. As she labored on the illustrations, Mary later wrote, she and Louis fell "deeply in love" (M. Leakey, 1984, p. 40).

## NOTES

1. Cutler never returned to England. He spoke enough Kiswahili to remain in Kenya after Leakey left but became weakened with malaria and died of blackwater fever on August 21, 1925.

2. In *White African*, Louis describes this staged incident as a Maasai raid (p. 190); in *Ancestral Passions* (p. 41), Virginia Morell reports that Elisabeth Kitson wrote a letter home to her parents in which she described it as a second lion roaring. Morell also notes the discrepancy. Additionally, Leakey spells Kitson's name as Elizabeth (p. 200) and Morell as Elisabeth (pp. 39–45, 52).

3. In *Ancestral Passions* (p. 45), Virginia Morell notes Louis's attributing the finds to Solomon and Kitson in *White African* but in the February 1965 issue of *National Geographic* magazine and in a guide published by the National Museums of Kenya, the finds are attributed to Leakey himself.

# CHAPTER 3

# POSITIVE CHOICES, NEGATIVE VOICES

*(Mary, 1913–1936; Louis, 1933–1936)*

In conclusion I wish to thank Miss Mary Nicol for her excellent drawings of stone tools which illustrate the book, and for her help in preparation of the map and index.

—L.S.B. Leakey, preface to *Stone Age Africa*

## FAMILY CONNECTIONS

The woman who became Mary Leakey, renowned paleo-anthropologist, first female recipient of Sweden's Linnaeus Medal, with honorary doctoral degrees from Yale University in the United States, Oxford and Cambridge Universities in England, and the University of Witwatersrand in South Africa, plus many other honors and awards, never graduated from high school. She was born on February 6, 1913, in London, England, and had a lifelong love of learning and a lifelong dislike of formal schooling.

On her mother's side, Mary was descended from John Frere, an eighteenth-century archaeologist who, in 1797, rightly proposed that flints he found in Suffolk, England, belonged to a remote period of prehistory. Members of the Society of Antiquaries, like most Christians at that time, accepted a literal interpretation of biblical creation and ignored his discovery. Other Frere family members were noted for taking equally unpopular views, including fighting against slavery.

Mary's grandfather, Douglas Frere, died from tuberculosis. He was still young, but unfortunately had already gambled away the family wealth. Mary's grandmother, Cecilia, was forced to sell her jewelry and the family villa in Italy and to rear her four daughters, one of whom was Mary's mother, in refined poverty in London.

Mary's other grandfather, Erskine Nicol, was an artist whose second wife, Margaret, was Mary's grandmother. Both of these grandparents died before Mary's birth and she was not close to Nicol cousins.

Mary's father, named Erskine Nicol after his father, was also an artist, and her mother, Cecilia Frere, was named Cecilia after her mother. The pair met in Egypt. Cecilia's high interest but limited talent prevented art from becoming more than a hobby for her, but it did provide the basis for a friendship that turned to love. After living on a houseboat in Nice, France, the couple returned to England for Mary's birth. Mary inherited only her middle name from her mother's father and was christened Mary Douglas Nicol.

World War I kept the family in England during Mary's younger years. After the war, in the winter months Erskine painted as he traveled with his family across Europe and northern Africa; when summer approached and money was low, they returned to London to sell his work. There the three lived with Cecilia's three unmarried sisters and mother in London. The sisters—Mary's aunts—were attractive but not, according to Mary, as attractive as her mother. Mary adored her grandmother, who advertised her heritage on St. Patrick's Day by singing Irish songs loudly while walking down King's Road in London with Mary in tow. Mary was shy and preferred more routine shopping trips with her grandmother.

Although the Frere family lacked money, her grandmother and three aunts had an abundance of love for Mary and enjoyed spoiling her. One Christmas they managed to give Mary the gift of her dreams, a doll with real hair who arrived with clothes made by her aunts—six outfits, including an evening dress trimmed in gold. Mary was overwhelmed with the clothes, "more lavish" (M. Leakey, 1984, p. 19) than any she could imagine. She suggested in her memoirs she may have inherited two other interests, painting and pets. All the Freres loved animals, especially dogs, and Mary's first attachment was to Jock, her aunts' fox terrier. Her first real pet was a cat that she nursed back to health while her father painted in France and she was heartbroken when she had leave it behind when they moved.

Although she was fond of animals, Mary cared little for wild or domesticated birds well into her adult life and later suspected she was wary because she associated birds with Aunt Bessie, her father's sister. Aunt Bessie loved birds, especially an Australian cockatoo. What she often didn't like was Mary's behavior, and to make her conform, Aunt Bessie threatened to destroy Mary's beloved stuffed bear, Pimpy. Aunt Bessie and birds, however, were minor irritations in a pleasant childhood, as Mary happily toggled between a stable life with her grandmother and aunts and an unpredictable nomad life with her parents.

The family's lifestyle offered opportunities for adventure. Her father's pleasure in "everything archaeological" (M. Leakey, 1984, p. 24) led them to investigate Paleolithic sites. Mary remembered watching Monsieur Peyrony, a specialist in French prehistory, sort through a bucket of dirt, select pieces that visually looked promising, and dump the rest into the river, a process that excited her at the time, when she was too young to wonder about the treasures lost to the muddy river bottom. Mary and her father searched through remains from M. Peyrony's excavation and retrieved all types of tools and blades from the Upper Paleolithic period, about 40,000 to 14,000 years BCE. Her mother's fluency in languages made her valuable as an interpreter on their excursions, but she did not share the enthusiasm of the other two family members. Mary and her father truly took pleasure in collecting and sorting their finds.

They also discovered cave art not yet open to the public when the priest at the village Catholic church, an amateur archaeologist, took them to see paintings Mary described as "magnificent" (M. Leakey, 1984, p. 28). The priest in the parish, Abbé Lemozi, was also a friend whose company Mary and her family enjoyed and with whom they went on long walks into the countryside.

As Mary moved into her early teen years, she and her father shared mutual loves of painting and prehistory and enjoyed each other's company. Change came rapidly in the spring of 1926, when she was thirteen. Her father, whom she considered "the best person in the world" (M. Leakey, 1984, p. 29), died at age fifty-eight from what was assumed to be cancer. Although she had always associated primarily with adults, childhood was quickly left behind.

## SCHOOLING, OR LACK OF

Like many young women in their teen years, Mary and her mother frequently clashed and her father's death exacerbated their problems. Adding to personality conflicts were money difficulties, as Mary's father had been ill during the winter and spring when he normally painted canvases whose sale would financially tide them over during the next winter; however, selling all his paintings earned enough that mother and daughter were able to move from Mary's grandmother's house to a boarding house.

Mary admitted that the major share of the blame rested with her for disagreements with her mother. Even when her father was alive, a very unattractive governess they hired to tutor her gave up after suffering much abuse from Mary and an American friend—they dubbed the poor young woman "The Uncooked Dumpling." (M. Leakey, 1984, p. 26). After her father died, her mother decided Mary needed a convent education, but, no matter what religious order directed the school, the results were disastrous. When Mary consistently cut her poetry class, punishment was to recite poetry to the entire student body, which Mary refused to do. She was expelled. At the next school, the nuns recognized soap, not a seizure, as the cause of Mary's foaming at the mouth. Mary was also guilty of setting off an explosion in the chemistry lab. She was again expelled.

After several tutors failed to motivate Mary to learn, her mother, who Mary wrote always acted the lady in spite of their problems dealing with each other, gave up. So, in 1930, at the age of seventeen, Mary began to make her own decisions. She became one of the first women in England to fly a glider and stubbornly decided to manage her own education and learn only what she dedicated herself to studying.

Mary continued to draw and also to informally attend lectures in archaeology when they were available in London. She was anxious to volunteer for fieldwork but, not wanting to confront archaeologists in person, introverted Mary wrote letters in which she volunteered to work at excavations in England. Beginning in 1930, she worked at Hembury in Devon for several summers and added other experiences in the field to what she learned through lectures.


## LIVES COLLIDE

Mary's career combination of artist and archaeologist prompted Dr. Gertrude Caton-Thompson to recommend her work to Louis Leakey and arrange for them to meet at dinner after Louis spoke at the Royal Anthropological Institute. Although in *Leakey's Luck* Sonia Cole described Mary as immediately "bowled over by his charm" (p. 106), Mary wrote that Louis asked only for her help in illustrating *Adam's Ancestors* and she was initially more intrigued with the challenge of artistically capturing the texture of the rocks than with Louis himself. Correspondence was friendly and dealt with the illustrations throughout the summer of 1933. Then, when they met by coincidence at a conference of the British Association for the Advancement of Science, their personal relationship quickly advanced too. Although Mary noted that all was proper at the conference, they both realized their futures would include each other.

Their nonprofessional relationship was soon obvious to friends and colleagues at Cambridge and the British Museum in Bloomsbury where Mary was working on some illustrations for Louis's later books. Louis made no secret of his affair with Mary and his moving toward divorce and remarriage. In the 1930s not only did most people generally oppose divorce, but those in Louis's social group were fond of Frida. Many mutual friends shunned both Louis and Mary, although several ignored the situation or quietly tried to convince Mary she was taking the wrong path. Just as she resisted a convent education, Mary stubbornly resisted pressure to break off her relationship. Sir Thomas Kendrick, at the British Museum, warned her about the brilliant but hyperactive Louis, "'Genius is akin to madness, Mary: you must be careful'" (M. Leakey, 1984, p. 43). Her mother felt Louis was, if not crazy, then unstable, and was furious at her daughter for getting involved with a married man ten years Mary's senior and the father of two small children. Even Mary's traditional allies, her doting aunts, were against the relationship and kept it a secret from Mary's grandmother. Most family members and friends eventually accepted her decision, except Gertrude Caton-Thompson, who felt Mary had betrayed her, and not until forty years later, after Louis's death, was the friendship repaired.

Louis was generally ostracized and suffered the greatest effects from the affair both personally and professionally. He lost his academic position at Cambridge, one that he loved, when his marital situation became common knowledge. Together Mary and Louis endured Frida's angry outburst in which she clearly let them know that she wanted both of them out of her sight and life. This meant that Louis could not see his children again. Even with the many negative voices they heard, both Louis and Mary were positively sure of their love and moved toward a life with each other.

## EAST AFRICAN ARCHAEOLOGICAL EXPEDITION #4

In spite of his personal problems, Louis managed to raise money for a fourth expedition, the last such expedition he would lead, and left for East Africa in October 1934. The first part of Louis's work on this trip was to be devoted to confirming the brilliant hominin discoveries he had made at Kanam and Kanjera in 1932. Geologist Percy Boswell was scheduled to visit the site to corroborate the discoveries and confirm the date of Louis's Kanam mandible. Some scientists, like Boswell, felt that the fossils had been resituated because of soil erosion and were not convinced Louis had dated them correctly.

Although he erred when he promoted Reck's Olduvai Skeleton as half a million years old, Louis received generally wide approval from colleagues for the Kanam and Kanjera finds. One of the few scientists who expressed doubt was Percy Boswell, a no-nonsense, meticulous geologist from the Imperial College of Science in London. Boswell was vocal about his misgivings and, at Louis's invitation, agreed to travel to East Africa to confirm or refute the finds. Boswell was dubious of Louis's claims that early humans came out of Africa, because, after all, Piltdown Man offered proof that England was the birthplace of early humans.

## OUT OF AFRICA? OR OUT OF ENGLAND?

Why was it so difficult to believe that the origins of humankind began in Africa? The skull of Piltdown Man was primarily responsible, although racism played a role. Westerners believed that culturally they represented the pinnacle of civilization and easily accepted Piltdown Man as evidence that the beginning of modern humans was in England. Except Piltdown was a hoax, a hoax that went undiscovered for fifty years and has to rank as one of the top practical jokes of all time.

In 1912 the time and circumstances were ideal for the Piltdown skull to appear and be accepted by both scientists and the public. Science had captured the public's imagination in the Western world, with Darwin's theories on evolution widely accepted after publication about fifty years before, about the time of the American Civil War. The Western world, reverberating with discoveries and inventions, had joined the calendar in racing into the twentieth century. Gregor Mendel's work with peas validated the theory of heredity, and in the Neander Valley of Germany, an early species of hominin that became known as Neanderthals had been found. Where would the "missing link," the bones that provided a connection with ancient humans, be found?

The British people, including scientists, were basking in the glow of a sun that, quite literally, never set on their powerful empire. Since Stone Age tools had been unearthed in their country, they were ready to believe that the missing link would logically be found there. The Piltdown skull with its big brain looked exactly the way people thought the connecting link between ape and man should look and, amazingly, few hard questions were asked about the circumstances surrounding the skull's discovery in a gravel pit or about Charles Dawson, the amateur fossil hunter who found it. In his honor, the skull was given the name *Eoanthropus dawson*, "Dawson's Dawn Man" (http://www.museumof hoaxes.com/piltdown.html).

Theories abound about the person or people who placed a skull, expertly modified by linking the jaw of a modern human with an orangutan's jawbone whose teeth were filed to match the skull's human teeth. When world-renowned British scientists joined the public in proclaiming that human life had indeed originated in England, lesser authorities didn't express their doubts, although various American anthropologists voiced skepticism. Leakey, to his credit, was accepting but not enthusiastic about Piltdown Man even though his supporter and mentor, Sir Arthur Keith, was England's most eminent scientist and one of Piltdown's strongest advocates. Louis's lack of interest may have been partly because he was so involved in the search for earliest ancestors in Africa.

Two world wars came and went until modern dating techniques exposed the hoax in 1953. A team of scientists at the British Museum's Natural History Branch, later renamed the Natural History Museum, first untangled the truth. One of the prime players in the investigation that revealed the deception was Kenneth Oakley, an eminent geologist with whom Mary Leakey wrote her first published scientific paper. Oakley, Wilfred Le Gros Clark, Joseph Weiner, and their team announced that the ape had died not long before its jawbone was attached to a human skull that was only 50,000 years old; later tests proved the skull was less than 1,000 years old. To this day, the riddle of who perpetrated the hoax remains unanswered. For many years, however, Piltdown Man was the point of reference for all major prehistory finds and was in 1935 when Boswell arrived to view Louis's Kanam jawbone.

## ONE MISTAKE AFTER ANOTHER

Louis arrived several weeks before Boswell, and nothing went right from the time he set foot on East African soil in the fall of 1934. At Kanam, the iron stakes he planted to identify the site where he had found the skull were gone, probably taken by tribal fishermen who used them to make fishing harpoons and spears. Louis and a friend had taken photographs but his turned out blank and the friend photographed only the general area. Nor did the location look familiar. In the intervening three years trees were cut down and heavy rains caused gullies so that Louis now saw a very different landscape. With his usual optimism, Louis was sure he could control events and prove to the world that Kanam man deserved to be linked to modern humans.

The situation was further complicated by similar problems at the

Kanjera site. Finances prevented Louis from having a surveyor on his previous expeditions and he paid dearly for not making even a rough sketch of the site where the fossils were found; pointing to a general area would not sway a meticulous geologist whose life revolved around seeing convincing proof. In the 1930s, dating techniques were in their infancy and determining the age of a fossil was primarily by the expert and experienced eyeball approach of qualified scientists who, like Boswell, read the soil.

Since propeller planes in the mid-1930s couldn't fly long trips non-stop, Boswell spent five days hopscotching south from England to Europe to Africa. He arrived on November 21, 1934, in a good humor, and interactions were initially quite friendly between him and Louis.

Louis, eager to prove to Boswell that his error in supporting the age of Reck's skeleton was an isolated case, enthusiastically shepherded Boswell from site to site, including Olduvai. Louis had been honest about his difficulties in finding the original locations of the Kanam and Kanjera fossils, but perhaps he also hoped to buy time by carting Boswell over hill and through Maasai village. Sam White, a surveyor, Peter Kent, a geologist, Juma Gitau, who had found the Kanam mandible, Heselon Mukiri, a Kikuyu with a terrific bent for fossil finding, Peter Bell, a zoologist, and Allen Turner, an anthropologist, along with other camp workers, had immediately set to work to locate fossils that would support Louis's claims, but the fossils they found were of little value in proving the location of the disputed Kanam-Kanjera finds.

While Heselon Mukiri continued an unsuccessful search with some members of his team, Louis took Boswell over dusty roads under Spartan conditions to visit sites not directly related to his long trip from England. Crisscrossing the country was wearing on Boswell. Herds of giraffe and other wild game sometimes reduced travel speed to ten miles an hour. They dined on buzzard and tough warthog meat. Boswell was anxious to see the place with a magnificent setting that "defies description and cannot really be captured by photography" (L.S.B. Leakey, 1974, p. 37), whose fossil wealth he had heard about, a "veritable paradise for the prehistorian" (L.S.B. Leakey, 1966b, p. 254), but when they at last arrived, he was not enchanted with Olduvai's black volcanic dust that permeated every item of clothing and every pore of their bodies. Even Louis—a man who thrived on austere camp conditions—once wrote that at Olduvai "everything was dust, dust, dust!" (L.S.B. Leakey, 1966b, p. 256). The hot sun combined with dust to make bodies filthy and there was dust in bedding, dust floating on coffee or tea, dust in the mouth and nostrils, and dust on top of buttered bread. At dinner by lantern light, dust combined with swarms of insects to layer everything on its trip from hand to mouth.

The personalities of the two men eventually created problems. Louis jumped like a grasshopper to seek out every possible spot where fossils might be found; Boswell wanted to investigate every aspect of a limited number of fossils. He had arrived on November 21 and Christmas Day came and went and Louis was still not working at the Kanam and Kanjera sites. On December 28 Boswell insisted they visit the sites he traveled so far to investigate. At last, in mid-January 1935, after two months of mistaken trails and many trials, Boswell left, feeling his trip was a waste of time.

Louis was depressed and feverish from malaria but thought he and Boswell had an understanding. He expected Boswell to wait until Louis returned to England and could respond personally to Boswell's criticisms, but the two men miscommunicated. Far from holding his silence, Boswell was excruciatingly critical of Louis in reports at various meetings and in an article that detailed the Kanam-Kanjera fiasco in *Nature*, Britain's most prestigious scientific journal. Louis, unaware of the criticism that could effectively ruin his career, had gone on to other projects and was shocked when mentor and friend A. C. Haddon sent a letter that alerted him to the problems caused by Boswell's condemnation. Boswell neglected to mention some of the reasons, such as the blank photographs, and Louis responded in the worse possible way. He criticized Boswell in a field report that sounded like sour grapes because it carped at Boswell for breaking his promise to maintain silence until Louis was present to respond. Eventually Louis retracted his statement, but it was too late to salvage a situation bad from the beginning.

Louis was unwavering in his belief that the Kanjera fragments were early *Homo sapiens*. In 1936 Louis wrote in *Stone Age Africa* that he expected "the whole evidence will probably be rediscussed soon at a conference" (p. 165) and that information would be included in the appendix if it were available in time. His tone assumes a finding in his favor. On the last page of *By the Evidence*, signed September 30, 1972, the day before his death, Louis wrote that he never "had the slightest doubt as to the validity of my Kanam jaw of 1931." Son Richard's recent finds, he went on to say, were "clear confirmation of the existence of men similar to those represented by the Kanam jaw in the earliest Pleistocene" (L.S.B. Leakey, 1974, p. 257). In spite of Louis's conviction, the Kanjera and Kanam specimens have been found to be comparatively recent (Roger Lewin, 2002, p. 86).

The damage to Louis's reputation was enormous and as long lasting as his outspoken belief in the Kanam jaw. Ironically, Boswell placed his faith in the fake Piltdown skull, even though little information was available about the date and exact location where an amateur fossil hunter found it. He discredited the fossil found by Juma Gitau because Gitau was only a Kikuyu "boy," not a real scientist (Morell, 1995, p. 83).

Louis Leakey had enjoyed enormous success at a much younger age than most anthropologists but, as 1935 rolled toward mid-year, his future was tarnished. Most of his colleagues had deserted him. The glow that surrounded him began to dull when he made serious professional errors in judgment. These errors were exacerbated by what many saw as grating confidence bordering on arrogance. And to some his audacity in leaving his wife and two small children and living with another woman darkened his future even further. But a major source of his difficulties was soon to speed recovery of his reputation. Enter Mary Nicol.

## THE LURE OF AFRICA

While Boswell was huffily returning to England and Louis was juggling the resulting fallout, Mary spent from January til April 1935 traveling with her mother in South Africa and Zimbabwe (then Southern Rhodesia). Her mother was hopeful that the distance between Mary and Louis would cool their relationship, but Mary, eager to discover the anticipated archaeological riches of Olduvai for herself, parted company with her mother, who returned to England while Mary traveled north to meet Louis. Olduvai Gorge, stretching thirty miles across the

Serengeti plains in northern Tanzania, immediately enthralled Mary and she later wrote that she would never tire of that view. "It is always the same, and always different" (M. Leakey, 1984, p. 55).

Mary soon proved her worth by being a team player and by showing she had expertise retrieving fossils from the field, but an expedition in Kenya was far different from excavations in England.

## CREATING A CAMP

Creating a camp in the early days was a challenge. Just reaching Olduvai was a trial. Mary wrote that when she first heard Louis speak he lectured about "that remote place Olduvai Gorge, which he had succeeded in reaching" (M. Leakey, 1984, p. 41), a huge understatement of what was involved in navigating the drive with three one-and-one-half-ton trucks and a car. The members of the expedition spent an entire day traveling about one hundred miles over dusty roads, and then the trucks' shocks strained following a track used occasionally by traders. On the last leg of the trip they crept at less than five miles per hour. Time was spent cooling down the engines and four gallons of precious $H_2O$ were consumed replacing water that boiled away.

Setting up a camp required planning far ahead to assure adequate food and water for the crew. Sometimes water for camp use was limited or, as a dry spell continued, almost nonexistent. It usually had to be brought by car or truck over bad roads or, worse, carried by foot. Both methods were tedious and labor and time intensive. Toward the end of the rainy season water was in short supply, and everyone in camp was forced to drink water in which rhinos had wallowed. Even filtering couldn't make tea that didn't taste tinged with rhino urine. After a surprise storm, they forgot their tents had been treated with insecticide and gleefully collected run-off water. The resulting drink didn't kill anyone but was noxious enough to cause everyone to be terribly sick.

Mary gamely ate the meat Louis was sometimes able to shoot, but her love for animals made dining on them at dinner difficult. When a truck was more than a week overdue, they existed only on rice, sardines, and apricot jam, a combination Mary found so revolting that she and Louis decided to drive by car to purchase supplies. When Louis drove into a gully, they had to sleep in the car and then spent the next day digging its wheels out—with table knives and several sturdy dinner plates. Some Maasai warriors, highly amused, watched but considered helping undignified. The truck arrived with food and towlines just as Mary and Louis completed the digging.

With no easily available means of replenishing supplies, every item served multiple uses. Louis wrote that gasoline was shipped to camp in cans stored in wooden boxes. The latter were then used for storage. When camp broke up, the team carefully wrapped collected fossils and placed them in the wooden boxes and the Maasai eagerly accepted the gasoline cans.

To Mary's delight, after their work at Olduvai, she and Louis spent three weeks camping at Kisese where Louis had promised to show her rock paintings. There fresh fruit, vegetables, and water were unlimited and they could even take hot baths. Then it was on to Nairobi for the trip back to England. But first Louis visited his parents.

Harry and Mary Leakey posed a major problem. Both were respected missionaries. Louis's father, an Anglican canon, and his mother felt their religion deeply and personally. They had already heard hints about his separation from Frida and relationship with a woman living in the camp. Exactly what Louis told Harry and Mary is not known, but he provided his parents with limited information about his relationship with Mary and, in fact, while he visited his parents, he arranged for her to stay at a Nairobi hotel. Despite this sensitive situation and the difficulties of camp life, Africa had, she wrote, "cast its spell" on Mary Nicol (M. Leakey, 1984, p. 63).


## MONEY MATTERS

Back in England in the fall of 1935, Mary was very happy at Steen Cottage in Hertfordshire where she and Louis lived almost a year. Local folks didn't know they were "living in sin," and Mary wrote that the villagers probably wouldn't have cared, but Louis was thrown out of the tennis club in the larger town of Ware when news of his divorce spread. Her mother and aunts sadly accepted the situation and Mary remembered only one critical comment from her Uncle Percy and it didn't pertain to her romantic situation—she hadn't tied up the dahlias in her flower garden correctly.

Since the couple had almost no money, their cottage had no heat, no electricity, and no water. And there were no job offers, although Louis cast a wide net in England, Kenya, Tanganyika, and even Rhodesia. Finally, early in 1936 he accepted an advance to write his autobiography. He considered *White African* "simply and solely what is often called a 'potboiler'" (L.S.B. Leakey, 1974, p. 73), written to pay bills. But it wasn't a potboiler by any means and earned excellent reviews.

Needing money did not make Louis more cautious about antagonizing readers. *Kenya: Contrasts and Problems* was a short book he wrote aboard ship in September 1935. White settlers were angered by his listing reasons the indigenous people hated and distrusted them. Other criticisms equally infuriated various groups of readers. He discussed why civil servants with limited funding could not carry out governmental policies that were inherently ineffective and discussed pros and cons of the work of missionaries. He wrote that white Kenyans should cooperate rather than try to dominate the country because "even though I am certain that Kenya will never be the white man's country in the sense that many would wish, I am certain that the white race has got its place, and that place can only be made safe and economically sound if the interests of the black man are studied to the full" (L.S.B. Leakey, 1966a, p. 183). This conclusion was the final insult for white readers, although a rational review of his charges and consideration of changes would have been in Great Britain's best interests, given the later struggle toward independence. (See chapter 5 for more on Mau Mau and Kenya's gaining independence from Great Britain.)

Louis also wrote *Stone Age Africa* after giving a series of Munro lectures at Edinburgh University in 1936. He was one of the youngest lecturers selected to present and, although being invited was an honor, he and Mary much needed the financial remuneration. Then, when the Rhodes Trust offered Louis the opportunity to study and write about Kikuyu history and traditions, he and Mary immediately accepted and then set about deciding how to make it happen. In *Disclosing the Past*, Mary described their philosophy as "not to fret over difficulties" (p. 67), but to take advantage of opportunities and make decisions as they encountered hurtles.

## UNDOING AND REDOING WITH LITTLE ADO

Frida had filed for divorce early in 1936 and finally, on December 24, 1936, Louis and Mary were married in London. Mary's mother, who eventually adjusted to Louis's divorce and subsequent status as son-in-law, attended the ceremony along with one aunt. The best man was Peter Mbiyu Koinange, son of the Kikuyu senior chief. He was staying with Louis and Mary but didn't realize they weren't already married until he was pressed into service. Mary prepared a lunch that included an excellent apple tart, and only at the last minute did they notice that pooch Bungey had chewed a large hole in Louis's coat. It was, Mary later wrote (1984, p. 67), about as far as possible from the romantic and fashionable wedding that mothers dream their daughters will have.

## CHANGING THE FACE OF PALEOANTHROPOLOGY

In January 1937, Louis and Mary returned to East Africa as a married couple. At this time the world's foremost anthropologists generally believed:

> Humans evolved in Europe or Asia.
> A simple straight line would go from earliest to modern humans.
> Humans had recent ancestry.
> First, the brain expanded as humans evolved, and then humans eventually walked upright (personal communication, Louise Leakey, October 16, 2003).

By the start of the twentieth-first century and through the work of the Leakeys and other anthropologists, each of these beliefs was disproved. But not easily and not quickly.

# CHAPTER 4

# WAR AND CHILDREN

### *Real Time (1937–1946)*

> One part of our research at Olorgesailie is the study of narrow bands of sediment that reflect ancient landscapes over which plants grew, animals lived and died, and early humans dropped their "Stone-Age business cards"—stone tools and other signs of their activity.
>
> —Dr. Rick Potts, Smithsonian Institution,
> personal communication, October 26, 2004

## WAR CLOUDS THE FUTURE

At the beginning of 1937, when Louis and Mary returned to Kenya, war clouds hovered on the horizon. Italy annexed Ethiopia, Kenya's northern neighbor, in May 1936 and controlled the country until 1941, but the Italians, who joined Japan and Germany to become the Axis powers, never attempted to move further into Kenya than the border country, and Mary wrote that during the years when World War II conflicted most of the world, she and Louis "were able to get quite a lot of archaeology of one sort or another done" (M. Leakey, 1984, p. 77).

Louis was anxious to quickly begin work on the history of the Kikuyu and record their traditions, customs, and language before the intrusion of Western ways into their lives. He immediately sought help for his project from Chief Koinange, the friend whose son served as best man at his wedding. Chief Koinange lived in a stone house rather than a traditional tribal home, had adopted Western clothes, and worked with British officials, but he was also sensitive to Kikuyu concerns and needs.

Because he was white, Louis's motives were suspect, even though he was himself an elder and could not have gained tribal support without that status. After a week of discussion and consideration, nine elders were appointed to advise and provide information to Louis. As with other projects he undertook, Louis soon immersed himself in comprehensive study, examining every aspect of Kikuyu life and culture. Never one to be without words and forging ahead with an incredible 8,000 handwritten words per day, within three months he had written five hundred pages and anticipated they were only half of the first of three volumes. Louis finished volume three in February 1939 but refused to edit the massive 700,000 words to a size the Rhodes Trust was willing to invest in publication. Academic Press, with support from the L.S.B. Leakey Foundation, eventually published *The Southern Kikuyu before 1903* in 1977, five years after his death. Gladys Leakey Beecher and Jean Ensminger helped with the final editing of the book. Beecher was Louis's sister who married into a missionary family and became a missionary herself; Jean Ensminger lived in Kenya and knew and worked with Louis on the manuscript previously (Clark, 1989, p. 397). Mary was pleased with its publication and felt it represented the love and understanding Louis had for the Kikuyu and their culture.

With Louis heavily involved in his project, Mary wanted work of her own and began a dig. She was suddenly ill with a particularly virulent version of pneumonia, a deadly disease in a pre-penicillin world. Her condition was so grave that her mother made the four-day flight from London to Nairobi. By August 1937 Mary was fully recovered and gave credit to Louis for passing on to her the will to live. The good news was that Louis's parents had initially felt awkward at the sudden arrival of a second wife ten years Louis's junior, but she was accepted into the extended Leakey family by the time she was well and Cecilia enjoyed knowing Harry and Mary.

When Louis was far enough along in drafting his Kikuyu history so that he needed only limited help from two of the Kikuyu elders, he and Mary moved north of Nairobi to Hyrax Hill, so named because the small rabbitlike animals lived there in crevices in the rocks. To save money, Louis and Mary and the workers lived in tents but built a grass hut to serve as a workroom. Despite a cobra that roomed in their roof, Mary described it as "very pleasant" (1984, p. 71).

## MARY'S EXCAVATION PROJECT

Hyrax Hill was completely Mary's project. She found indications of settlements dating from 200 years past to the Neolithic, or New Stone Age, period, about 2,000 years in the past. In addition to stone implements, pottery, beads, and human bones, she found ancient cup marks made in the rocks that she believed were used for the mathematical strategy game of *bau*, still played around the world. The game goes by a huge variety of names, including *Mancala*, with variations of rules and equipment—like pebbles, beads, or seeds—with which to play. She couldn't date the cups and did speculate that the game has prehistoric roots.

Her work at Hyrax Hill became such a local attraction that Mary opened it to visitors and was delighted when over £100 was contributed; a newspaper fund added more. A neighbor who supported Mary's archaeological work purchased the property and gave it to the government and today Hyrax Hill Prehistoric Site is one of the National Museums of Kenya.

During the 1937 Christmas holiday, Nellie Grant, a strong, independent woman who became a friend, introduced them to the Njoro River Cave, a Late Stone Age burial site where Mary, Louis, and a small team eventually found about eighty bodies, each with a stone mortar, pestle, and other artifacts. They estimated the burials dated to about 850 BCE. Later radiocarbon dating proved they were remarkably accurate when the date obtained was 970 BCE. plus or minus eighty years.

Although contributions helped with excavation expenses, the couple constantly hurt for money. Nellie Grant wrote a letter in which she described Louis and Mary as existing "'on the smell of an oil-rag'" (M. Leakey, 1984, p. 75). Their situation became desperate in 1939 when Louis's grant for the Kikuyu project ended. By this time the war clouds were so dark that returning to England was impossible. They bought only necessities and survived on what money Louis earned with two minor jobs. In the first, he bought goods wholesale in Nairobi and trekked to small Kikuyu villages to sell them retail. The traveling salesman job was a cover for his second job of intelligence work for the British government in tracking down the source of war propaganda spread among the Kikuyu.

## EAST AFRICA DURING WWII

World War II began in Europe on September 1, 1939, when the Nazis swept across Poland, followed two days later by Great Britain's declaring war on Germany. In the early 1900s Germany ruled over a colonial empire in Africa, and, to protect its interests, fighting was fierce and greatly impacted life in Kenya during World War I. When Germany lost the war, it lost these African colonies, with the result that World War II touched African lives minimally because action was in northern Africa; however, goods like gasoline, needed for war machinery, were in short supply.

Although gas rationing prohibited extended fieldwork, Louis had time during his official government travels to search for likely sites in Kenya and Ethiopia for future expeditions. He found many sites that reeked of prehistory "high potential" and described not being able to investigate them as "most tantalizing" (L.S.B. Leakey, 1974, p. 168). Louis wrote little specifically about his wartime work but clearly relished an expanded role in intelligence that included running guns to Ethiopian guerrillas fighting the Italians. Such secret activities tapped into his theatrical side.

Louis also was responsible for surveillance of Italian prisoners-of-war who were captured in battlefields to the north and sent to Kenya. When several Italian prisoners attempted to escape, a young African volunteered to lead Louis and his men from the African intelligence service to some *azungu*, or Europeans, who, the young man said, badly needed help. They turned out to be the escapees and "were huddled, very cold and hungry" under a tree. City dwellers in Italy, they had been badly frightened by the horrible screams and screeches of a ferocious animal and begged "to be taken back, as quickly as possible, to the safety and comfort of the prisoner-of-war camp" (L.S.B. Leakey, 1974, p. 142). The ferocious beast with the piercing screech was a tree hyrax, small, furry, and harmless.

In 1940 Louis was made curator of the Coryndon Memorial Museum, so named in honor of Sir Robert Coryndon, a governor of Kenya with a keen interest in history. The position with the museum, which is now the flagship of the National Museums of Kenya, came with no salary but offered a much-needed perk. Their first child, Jonathan, had arrived on November 4, 1940, and they moved into a small house next to the museum early in 1941. One of Nairobi's first houses, it was made of now rusted corrugated iron, with holes that invited rats and mice. Second son Richard wrote in *One Life* that he could "clearly recall the rat hunts with the dogs on Sundays, when thirty or forty of these rodents might be killed" (R. Leakey, 1986, p. 17).

Insects also easily found their way into the house. When Jonathan was tiny, siafu ants climbed through holes in the walls and attacked him with their pinchers. Siafu is a Swahili word for ants and, though they're not large compared with others of their species, siafu travel in armies and are vicious flesh-eaters that cover a victim before all begin to attack at once. Luckily, Louis and Mary heard the screams and immediately rescued the baby and picked off what Louis wrote was "a seething mass of insects . . . clinging to his face, eyes, and other tender parts of his body" that "had to be taken off one by one, as carefully as possible" (L.S.B. Leakey, 1974, p. 157). In spite of the house's disrepair, they were glad to live on the museum grounds, surrounded by fifteen acres of trees, exotic plants, and birds.

Although Mary wrote that she "quite liked having a baby," she followed it up by noting that she had "no intention of allowing motherhood to disrupt my work as an archaeologist" (M. Leakey, 1984, pp. 79–80). And she didn't. She much enjoyed work in the field and was well aware Louis had strayed from Frida when children interrupted his first wife's professional life. When Jonathan was five months old, Mary left him at home while she traveled with Peter and Joy Bally,[1] two botanist friends, and Jack Trevor, a physical anthropologist who had a small grant from Cambridge, to Ngorongoro to investigate burial mounds. A nurse cared for Jonathan during the day and Louis quickly learned to change diapers, prepare formula, and "bring up Jonathan's wind by holding him over my shoulder and patting his little back" (L.S.B. Leakey, 1974, p. 155).

## CARVED IN STONE

Louis and Mary made short fossil-hunting trips as Louis's leave and gas rationing permitted. On the Monday after Easter 1942, Louis and Mary, with several friends and members of the museum staff, drove their old car southwest toward Mt. Olorgesailie, an extinct volcano.

Although only forty miles from Nairobi, Olorgesailie is 2,000 feet below, where the Ngong Hills open into the wide expanse of the Great Rift Valley. The Great Rift is a crack in the earth's crust that meanders over 5,000 miles north to south through Africa. Lofty mountains and volcanoes frame the valley's edges and the dramatic countryside is dotted with lakes. The Rift Valley is a natural reservoir whose soil, through changes in climate and shifting of earth, offers up clues to early human origins. Fossils and artifacts buried under layer upon layer of soil for millions of years are moved upward as the earth continues to shift and finally water and wind expose them.

A brief description in a 1919 book about early tools found in the general area intrigued Louis and he was hopeful they would find some hand axes. What they found was their "most stupendous wartime archeological find" (Morell, 1995, p. 125).

The group fanned out over a promising area and Louis soon spotted some hand axes, but he had time only to mark the spot and then rush to Mary's call. She was staring with amazement at an unbelievably huge cache of hand axes. Mary later described it as "an extraordinary sight" (M. Leakey, 1984, p. 82), with hundreds of hand axes, all looking as if they were recently abandoned there. Soon everyone in the party had found similar, though less numerous, troves of tools.

Hand axes are more than just a Stone Age version of Swiss Army knives, as Dr. Rick Potts, Director of the Human Origins Program at the Smithsonian Institution in Washington, DC, calls them. How they were used isn't known for sure but assumptions are made that, like a Swiss Army knife, they were a multipurpose tool for cutting and scraping animal hides, digging in the ground, hammering bones, and whatever other jobs could be done by a handy tool the size of a fist and pear-shaped with faceted edges. Unlike other tools early humans probably developed from materials such as vines and grasses that easily deteriorate, these stone tools still exist. They now have been found in numerous places around the world and are known generally as Acheulian tools, after St. Acheul in northern France, where ones made of flint were first described. (Older tools are known as Oldowan, for Olduvai, where they were initially found.) Other tools found include cleavers, or u-shaped tools with a flat edge; flakes chipped from the main rock that experts suspect were used for skinning; and spheroids, or round balls. The Olorgesailie tools are crafted from fourteen different types of local lava rock, but two of these types make up 56 percent of the tools (Rick Potts, personal communication, October 26, 2004). Early humans later refined the tools into slightly more sophisticated versions, but why upgrade when something works? The tools were the same basic size and shape for more than a million years.

In the 1940s and up till his 1974 posthumously published autobiography, Louis hypothesized that the round spheroids were used as killing tools by hurling them in groups of three at animals. After all, he reasoned, South American gauchos connect three bola balls of various sizes with rope, whirl them around their heads, and toss them to entangle an animal's legs and bring a large beast down. Even in the 1940s Louis's theory convinced few people and in *By the Evidence* (p. 167), he wrote that he "nearly killed" himself when he experimented with hurling them. Archaeologists have since found these round stone balls at many sites and believe they're the result of continued chipping when making the hand axe itself. Referred to as hammerstones because they are used to hammer out chipping flakes to shape the hand axe, they are a size that fingers can neatly grip when giving a good whack to the stone held by the other hand. In other words, the hammerstone used in this production process becomes a rounded sphere from the continual whacking of the stone that becomes a hand axe.

Louis and Mary were eager to begin excavation when they first found the wealth of Olorgesailie in the spring of 1942, but they were unable to return for over a year. Louis was involved with the museum and intelligence work, and Mary was expecting their second child.

When they briefly revisited Olorgesailie in December 1942, Mary was due to deliver. Sadly, Deborah Leakey died of dysentery when she was three months old and neither Mary nor Louis wrote much of her death in their autobiographies. In fact, Mary slipped mention of Deborah in with a paragraph on the death of four Dalmatians, but friends reported that this was a time of great pain for them. Louis uncharacteristically lost his enthusiasm for work and later wrote apologetic letters for neglecting correspondence (Morell, 1995, pp. 126–27).

## OLORGESAILIE

Louis's old energy, stimulated by plans for excavations at Olorgesailie, gradually returned and in August 1943, during three weeks of leave, he and Mary set up a camp and began work. Louis came from Nairobi as often as possible evenings and weekends after his leave ended and was responsible for directing the dig there. Several older African workers not involved with the war effort helped, as did captured Italian soldiers delighted to spend their war years in the open air of Africa rather than being forced into the confines of a prisoner-of-war camp. A nanny, or *ayah*, was on site to care for Jonathan, then not yet two years old and a very accommodating, easy child to take along.

One important find was the skull of an extinct hippopotamus. Even the Maasai, not really enthusiastic about all this digging, agreed that the remains must be very old since no hippos could live in today's desert.

For the Leakeys, the best Olorgesailie offered was what they found on Easter Monday 1942—stone tools, but no fossil hominin bones to show who made them. By 1947 Louis and Mary stopped digging at Olorgesailie. Louis, who was masterful at negotiating with the locals, arranged with the Maasai for some of the land to be used as an open-air museum and built a raised catwalk around Mary's initial find of tools. The museum, now expanded and part of the National Museums of Kenya, provides visitors with an explanation of a significant step forward in human behavior. In 1947 Louis spiffed up the area to show to attendees at the Pan-African Congress of Prehistory and Palaeontology held in Nairobi, but then he and Mary moved on to other projects.

In 1994 Rick Potts, who was then conducting digs there, used some grant money to make the museum building a more permanent structure and rebuild the catwalk exactly as it had been. Just before Mary died in 1996 she told Potts she regretted that most of the hand axes had been removed or stolen because she and Louis had wanted people to see the area just as they had first seen it. Potts set up an old photograph of the catwalk and took Mary and daughter-in-law Meave to stand in the exact place where the photograph had been taken. When they compared placement of the stone tools with the photograph, every stone was there, in its original position. Potts laughs when he says it was the only time he remembers Mary saying she was wrong—and she was delighted to say so. The stones appeared different because the sun's heat caused them to develop a patina that changed their color.

Olorgesailie seemed to be simply a repository of early tools and faunal, or animal, remains. Admittedly, its hand axes made it what Louis wrote was "unquestionably the richest site" (L.S.B. Leakey, 1974, p. 185) ever found, but Mary did not publish any articles about this site that had initially thrilled them. Why, they must have asked themselves over and over, were there so many signs but no firm hominin evidence at Olorgesailie?

Mary and Louis were theoretically incorrect on several counts, which is a polite way of saying they decided to dig at the wrong place. First, they hoped the Olorgesailie site was what they called a "living floor," a term that today is considered imprecise and generic, and believed early people had made their homes along the bed of what, until about 180,000 years ago, was a lake. Picture this: The area flooded, these Paleolithic people moved on, and silt covered their early camp. Then, when the area dried out, they returned and rebuilt their camp. As this cycle repeated itself, various levels, or floors, built up, sort of like a geological pan of lasagna. Louis and Mary reasoned that human remains would eventually be found by digging through the layers.

Second, early humans may not have died there because they didn't live there. Potts theorizes that the lowlands by the streams at Olorgesailie were where early humans worked and spent their days, but no human fossils have been found there because they moved at night to higher ground. Since many animal predators stalk sources of water at night, remaining there would have invited death.

Every year since 1985 Potts and teams from the Smithsonian and the National Museums of Kenya have searched for Olorgesailie's hominin secrets. Based on Potts's theory that early humans lived—and died—on high ground, logically that is where excavation would take place, except that sediment is not deposited on high ground so remains would be washed to a lower level. The most logical place to search for remains of early humans, Potts suspected, is along the ridges formed between high ground and water's edge. For nineteen years Potts and his teams found what the Leakeys found—a fascinating array of thousands of hand axes and the remains of animals, including many now extinct, but elusive evidence of the existence of hunters who made the tools. At last, in 2003, sixty-one years after the Leakeys' initial visit to Olorgesailie, Potts and his crew held in their hands an adult cranium of the first hominin found there. It was excavated from a volcanic ridge. Dating back almost a million years ago and smaller than today's humans, so small it is presumed to be female, this portion of skull shows marks that suggest death from mauling by a large animal like a lion. She is affectionately referred to as "our little gal" (Nalencz, p. 26).

## PRIVY COUNSEL

Mary, Louis, and Jonathan took another brief wartime trip with friends to Rusinga Island, an island whose fossils included extinct species of apes. In 1943 the trip required a long bumpy overland drive and then several hours by boat across Lake Victoria. Louis arranged free overnight passage for the group, but they had to sleep rather uncomfortably atop sacks of grain on an Arab dhow, a vessel about forty-five feet long and driven by wind. Except that, partway there, the wind died down. Mary recalled the boat crawling with huge cockroaches but, even worse, the only lavatory was a seat over the side of the boat and, at daybreak, after some tea, bread and cheese, nature called. The crew was used to the arrangements but the Leakey party was not. They counseled with each other and arranged that two of them would hold up a blanket to make a screened privy. With little wind still, the captain pulled into a cove where they spent the day, grilled fresh fish for dinner, and finally arrived at Rusinga to begin their explorations. Mary remembered little being accomplished—but the trip was memorable. Successful fossil finding on Rusinga Island was in the future.

## A SECOND SON

Mary gave birth to her own second little guy, Richard Erskine Frere Leakey, on December 19, 1944.[2] World War II was ending and along with it Louis's government work. With two children, he needed a job. Except for use of the house, the Coryndon Memorial Museum refused to reward him for his work there and Louis felt he was unappreciated. And he was, though exactly why is not recorded. Admittedly, he angered some white members of the community when he insisted on opening the museum, unlike most public buildings in Nairobi, to Africans and Asians. Both he and Mary worked long hours for the museum, and he effectively organized speeches, films, and exhibits; responded to inquiries whether posted or in person; hired, trained, and directed staff, including Italian prisoners-of-war; and served as an educated and experienced anthropologist. After complaining to the trustees, Louis was given a stipend of £150 and a rather tart note stating that the money in no way obligated additional payment to be made for his services.

Louis was tempted by possible positions in England but the Coryndon eventually offered him a piddling salary, far less than his predecessor had earned as curator five years before, but at least sufficient wages to allow Louis and his family to remain in Kenya. When the health department declared their house on the museum grounds unfit for humans, the trustees moved him and his family to another home until a new one was built.

Louis and Mary realized they needed to return to England. Renters who wanted to purchase Steen Cottage had stored their personal possessions in a barn and Mary's mother was ill. In December 1945,[3] Mary, Jonathan, and Richard left for England by ship and Louis followed later by plane. London was a very different place from the city Mary had last seen eight years before. German bombing had devastated large areas and people had not recovered emotionally or materially from the war; many items of food and clothes were still in short supply. She was able to visit her mother, who died only two weeks after their arrival. Cecilia was buried at Assumption Convent, where Mary had endured a brief career as a schoolgirl.

Louis, as usual, had a variety of adventures on his trip. Flying by plane, he was quarantined in Cairo, Egypt, for twenty-four hours because his immunization for yellow fever was too recent for his certificate to be valid. Fellow passengers contacted the British Embassy and he was released from behind barbed wire confinement, only to be told planes to London were overbooked. Suddenly a woman debarking from a plane *en route* from India to England fell and broke her leg. Louis took her seat and "Thus I got passage to England in time to have Christmas with my family" (L.S.B. Leakey, 1974, p. 195).[4]

Louis, eager to begin planning the first Pan-African Congress of Prehistory and Palaeontology, took advantage of the trip to consult with old friends and colleagues and tended to as many organizational details as possible while in England and France. The politics of World War II were still raw and many prominent scientists felt their French counterparts had collaborated with the Germans during the war. Relationships were strained, but Louis was firmly convinced he must persuade several French scientists to attend, since they had pioneered excavations in northern Africa. He particularly wanted the Abbé Henri Breuil, the preeminent figure in prehistory, to preside over the congress and he agreed. Besides the usual problems and housekeeping arrangements involved, organizing such a multinational event presented exceptional difficulties. Many of the scientists were elderly, and events included visits to sites spread over East Africa. These side trips had to be coordinated and often required setting up remote camps with arrangements for eating and sleeping. Louis even hired a full-time mechanic and arranged for a "break-down vehicle" to go everywhere with the attendees, in case of emergency.

In addition to selling Steen Cottage while in England, Mary and Louis relaxed and visited old friends. Jonathan loved riding an escalator up and down in the departments stores and also enjoyed visiting the farm of a member of Mary's Hembury dig who had been supportive during the trying first months of Louis's and Mary's relationship.

Louis and Mary were now primed for a renewed push in their search for humankind's beginnings. Mary recalled the trip as a wonderful family vacation, but she and Louis also realized that their future "lay entirely in East Africa" (M. Leakey, 1984, p. 90).

## NOTES

1. After her second marriage, Joy Bally was well known as Joy Adamson and wrote several books, including *Born Free*, about the lioness Elsa.

2. Louis wrote in *By the Evidence* (p. 183) that Richard was a Christmas Eve arrival, but other references, including Richard in *One Life*, give December 19 as his birth date. December 1945 is the date Mary gave in *Disclosing the Past*, p. 88, and she described in detail the work she put into knitting scarves and sweaters for the boys and having overcoats made out of blankets because they were going to England at the coldest time of year and, not needing those items of clothing in Kenya, didn't own them. Sonia Cole, a friend of the Leakeys, also uses this date in *Leakey's Luck*. However, in *Ancestral Passions*, p. 137, Virginia Morell gives the date as early March 1946.

3. Again, the date of early July 1946 that Morell (*Ancestral Passions*, p. 140) gives for Louis's arrival is at odds with Louis's remembrance in *By the Evidence*.

# CHAPTER 5

# PEACE AND TURMOIL, PERSONAL AND POLITICAL

## (1947–1958)

Yet digging still with urgent finger
On fever-stricken isle Rusinga,
Where he and Mary oft did linger
In Kavirondo Gulf, no less,
And there Proconsul with finesse
Was lifted from lacustrine soil
To keep the family tree a-boil!

—Phillip Tobias's toast to Louis Leakey,
celebrating the 100th anniversary of his birth,
as cited by John G. Fleagle, 2004, p. 4

## A CAREER REVIVED AND RUSINGA ISLE

The January 1947 Pan-African Congress of Prehistory and Palaeontology was a huge success. The scientists left generally acknowledging that Darwin was correct about finding the roots of human origin in African soil. The conference not only helped move Louis's career back on track, but attendees recognized Mary's contributions to their partnership.

Interest in the Miocene period, from twenty-five to five million years ago, was high by the end of the Pan-African Congress, in great part thanks to Louis's enthusiasm. The geology of Rusinga Island in Lake Victoria dates from the early Miocene, with fossil specimens from eighteen to twenty million years ago, so after the Pan-African Congress, Kenneth Oakley, John Waechter, and Dorothea Bate stayed on to visit there. Mary called the trip an "arduous undertaking even under ordinary circumstances for the elderly Miss Bate" (M. Leakey, 1984, p. 94). Despite a lifetime flouting convention by traveling and exploring alone to satisfy her desire to learn about fields in which few women dared to tread, at age sixty-nine Dorothea Bate's morals were thoroughly Victorian. The group spent a day visiting several sites and then boarded a boat that an inexperienced captain promptly grounded. As the boat rocked and knocked, the good news was that they were not smashed to pieces, but the bad news was that the boat couldn't be freed. Islanders who saw their plight tried to help. "Soon," Louis wrote in *By the Evidence*, "there were some thirty naked men and a few women milling round in the water, desperately trying to get the boat afloat" (p. 210). Miss Bate kept her eyes closed tightly while Mary tried to comfort her. Alas, the last resort was to wade, which Miss Bate was not outfitted to do, or be carried to shore. Louis wrote that she was "absolutely covered with confusion and embarrassment" (p. 210). This was, of course, more than the Luo fisherman who hoisted her onto his shoulders was wearing. The group then had to trek seven miles across the island, with only one lantern, to their camp.

Louis and Mary decided their next major project was to investigate Rusinga Island. In addition to the post–Pan-African Congress trip, Louis had done some exploratory work there during expeditions in the 1930s. On Mary's first trip to Rusinga with Louis in 1942 (see chapter 4), he found part of a jaw he named *Proconsul nyanzae*. The latter word refers to the Swahili name for Lake Victoria, but why *Proconsul?* In the 1920s a trained chimpanzee named Consul amused spectators at London vaudeville shows by wearing a hat and jacket and smoking a pipe while he pedaled a small bicycle. When British scientist A. T. Hopwood found some fossil bones he believed were the relative of a chimp, he humorously selected *Proconsul* as the genus name, meaning that his fossil find predated Consul. In scientific circles, the first person to publicly introduce a fossil has the honor of naming it; hence, the name chosen by Hopwood belongs to that genus.

A brash American who was garnering huge amounts of money and equipment for a Kenyan expedition provided additional motivation to investigate Rusinga. Louis half-heartedly helped Wendell Phillips and his colleague, South African Basil Cooke, but not having received a grant for fourteen years, Louis was not about to risk failure and made it clear to Phillips that Rusinga was his territory. Louis's possessiveness was not uncommon among anthropologists then nor would it be now, as they often preface talking about the group or area they're investigating with "my," as "my site," and clear conventions exist about treading on another anthropologist's selected soil, dental pick in hand.

In January 1948, Charles Boise, a wealthy American businessperson living in London, read an account of interesting finds Louis made at Rusinga and offered financial help. Louis decided he had nothing to lose and asked for £1,000. Boise added a second check in the same amount when he heard about Phillips's plans and, these donations, coupled with a grant, sent Louis and Mary on their way to fossil-search more systematically on Rusinga Island in September 1948.

Human and prehuman fossils from long ago are not easy to find and those in excellent condition even more rare. First, since burials began only about 70,000 years ago, remains were never in a central burial site and, second, if a predator was involved, the body was scattered by scavengers over a wide area.

Conditions for accidental preservation also had to be right and rarely were; the body had to be protected after death, probably by sinking into a riverbed where the flesh and soft tissue decomposed slowly, undisturbed. The parts of the body that distinguish a living organism—hair, eyes, and skin—rapidly deteriorated, leaving only bones and teeth. Sometimes, however, these hard parts of the body were separated, so that a tooth washed far away from the jawbone to which it was once connected.

If good conditions continued over time, rain would wash sand and gravel from the riverbed to cover and protect the remaining body parts. Gradually minerals from the sediment that built up around the body replaced bone calcium and eventually, over hundreds and then thousands and finally millions of years, the bones become fossilized but remain hidden.

Now luck takes a new direction. Possibly the climate changes, the river dries up, and the earth shifts. Eventually rains helped the wind uncover the long hidden, now fossilized remains. Richard Leakey once described fossil hunting as "a long-term business because every year there could be something new" (1986, p. 24).

These early human or prehuman fossils are then found only if a paleontologist with an excellent eye and a degree of luck spots the fossilized remains.

## PUTTING THE PUZZLE PIECES IN PLACE

On their 1948 trip to Rusinga Island, Louis and Mary sensed that they might find something important. "The problem," Louis wrote, "was to locate it" (L.S.B. Leakey, 1974, p. 226). On October 2, Mary left Louis excavating a crocodile skull and walked slowly along the bare brown earth. Crocs, whether alive or long dead, were not what she cared about. As her eye moved up an incline, she noticed a tooth. She yelled for Louis and they realized the tooth was embedded in a jaw. The skull was fragmented, but it was obvious this was truly a find. Although fossil pieces of *Proconsul africanus* had been found, this was the first skull, and skulls provide an extraordinary amount of information.

News accounts make the discovery of a fossil like Mary's seem as if the specimen works its way to the earth's surface and is picked up or, with a little work, extracted easily from the soil. In reality, the work is tedious and time-consuming. Several days were required for excavation of the skull and then the surrounding soil was sieved, a process that involves systematically sifting sections of soil through a screen to be sure nothing is left behind. The dust falls easily through and any bone fragments are left on top, for careful sorting to reclaim each shard of bone turned to stone. Heselon Mukiri, Louis's lifelong friend and a competent and trusted field worker, carried out this job.

Once the pieces were collected, Mary began piecing the skull together. Although this was the first *Proconsul* skull found with teeth embedded in the jaw, there were still over thirty separate pieces, many of them tiny fragments, and the job required hours of intensive and concentrated work. Once Mary dropped a small flake of bone, crucial because it linked two larger pieces, onto the dusty tent floor and spent much time locating it.

Fast forward to the late 1960s, when Martin Pickford, a longtime acquaintance of son Richard and himself a keen-eyed paleontologist, happened upon a box in which fragments of small bony plates that are found under a tortoise shell had been stored since 1947. He detected, among the other remains, some hominin fragments. "Almost unbelievably," Mary wrote in her memoirs (1984, p. 99), Pickford had found pieces missing from the *Proconsul* skull.

## PERSONAL PEACE

Mary and Louis were delighted with their find and with each other. To personally celebrate finding the skull of *Proconsul africanus*, they decided to add another child to their family of Jonathan, now age eight, and Richard, almost four. They welcomed Philip nine months later, on June 21, 1949.

Professionally, they were anxious to show the skull to Le Gros Clark, an authority on primate evolution who had assisted with their obtaining a grant. Unfortunately, they were stranded on Rusinga Island until the boat came to retrieve them and it was mid-October before they arrived back in Nairobi. Even though Louis badly wanted to present *Proconsul* to Clark himself, he deferred to Mary, since she had found it. Always clever with organizing publicity feats, Louis persuaded British Overseas Airways Corporation to carry Mary, with *Proconsul* nestled in a box on her lap, to London in return for the positive press generated. Although later disproved, at the time this Miocene primate skull was thought to possibly be the "missing link" between apes and humans and everyone involved was very pleased. Mary was so pleased that she didn't mind the media attention, especially since it was focused on the skull and not on her. Le Gros Clark wrote an interim report immediately and it, along with media reports, encouraged the Kenyan government to provide grants that supported the Leakeys' work through the 1950s. Charles Boise wrote another generous check and the British Museum put the skull on exhibit in 1949.[1] *Proconsul* truly opened the path for the Leakeys' future work.

With Boise's financial assistance, Louis purchased a second-hand

44-foot cabin cruiser that he named *Miocene Lady*. Never again the frustration of being stranded with a fantastic fossil find on an island in Lake Victoria!

The family made a number of pleasurable and profitable trips to sites in the Lake Victoria area during the late 1940s and early 1950s; expeditions there lasted about a month and had a holiday aura. Luckily so, because Leakey vacations were almost always working vacations. Despite a sunrise start in the truck Boise's generosity purchased, the trip from their Nairobi home to the northeastern shores of Lake Victoria took all day. When the boys were older and small places to buy drinks popped up, Richard argued for buying the new, fizzy bottled drinks, but Louis preferred to make a roadside fire and boil water for tea. Another early start in darkness the next morning allowed them to arrive at Rusinga during Lake Victoria's cool morning calm. Luxury cottages today circle the tree under which the Leakeys picnicked and comfortable lounge chairs now face the waters of Lake Victoria. For the Leakeys the *Miocene Lady* was usually home for the family and the tents of visitors and workers dotted the lakefront, with one large mosquito-proofed mess tent where they also wrote up notes in the evening.

Today visitors who debark from their plane on the well-maintained air strip are welcomed by a tall Kenyan who hands them a cooling camphor cloth to wipe away perspiration; half a century ago visitors refreshed themselves only after Louis shot a gun into the water to scare crocodiles away. At Rusinga Jonathan and Richard followed Louis on his afternoon walks to find possible fossil sites and, along the way, learned about the island's natural world, not for exploration by the uninitiated. The boys were with their father when he stalked and shot a dangerous croc much feared by the villagers. While skinning the almost 16-foot-long beast that probably weighed a ton, they heard a rustling in the bush. Marching toward the carcass was a siafu ant army that smelled blood and the group escaped only by wading into the water. The next day they found the carcass, bones completely bare.

To the Leakey family, the benefits and pleasure of life at Rusinga and other sites outweighed the dangers and difficulties. Richard wrote about eating fish fresh from Lake Victoria for breakfast: "To taste pan-fried fish in the perfection of an African dawn must be one of the most delightful of all experiences" (1986, p. 23). At the other end of the scale was eating his father's kedgeree for breakfast. Louis enjoyed making this dish, but only Louis and Jonathan enjoyed eating it. Its ingredients included rice, chopped hard-boiled eggs, and canned sardines sautéed together, a combination Richard hated. Hated so much, in fact, that Mary remembered after having it for breakfast one morning, Richard stuck his fingers far down his throat "with predictable results" (1984, 108). Although Louis was angry, Mary was only annoyed and "felt secretly thankful that 'kedgeree' would probably not be served for breakfast again. Nor was it" (1983, p. 18).

*Proconsul* was their key fossil find on Rusinga Island but there were many others of importance, large mammals to small rodents—even an ancient ant colony complete with all the types of ants from eggs to soldiers and workers. All three boys learned fossil finding as toddlers and as children they "had more field experience under their little belts than other academics ever get in their entire careers" (Hellman, 1998, p. 166). When Richard was about six, his parents wearied of his whining about the heat and flies and sent him off to search for a fossil. Spotting a brown bone peeking out of the soil about thirty feet from his parents, he set to work with his own set of dental picks to dig it out. Soon Louis and Mary realized he had found the first complete jaw of an extinct giant pig. Richard was furious when they took over.

Although the family enjoyed working vacations at Rusinga and on another island, Mfangano, Louis and Mary were anxious to begin exploring Olduvai Gorge in Tanzania.

## OLDUVAI GORGE

In 1931 Louis first began exploratory work at Olduvai and Mary was mesmerized on her first trip in 1934 by its beauty and the likelihood of fossils. The years had slipped by and they had carried out little more than a general assessment of its potential, although Louis wrote a book, *Olduvai Gorge*, about stone tools he found there.

Olduvai Gorge was beautiful but also remote, rugged, and inaccessible. Travel was better than twenty years earlier but an extended expedition at Olduvai still required planning, perseverance, and ingenuity. The Leakeys spent Christmas 1950 there, making preliminary excavations, plotting possible sites on their map, and finding a shorter route for transporting water. Unlike Olorgesailie, the scarcity of water meant that the three boys were left behind with caregivers, although Richard remembers not understanding why they couldn't go.

When Charles Boise visited them in East Africa in 1951, he promised more substantial and continuing support. The trip proved to Boise how welcome his help was when, traveling from Olduvai to see rock paintings, eighty miles from a garage, a spring broke on the car Mary was driving. How to repair it? Louis sent someone to find, buy, kill, and skin a goat. Meanwhile, he and some of the others felled a tree and used the wood to make splints to hold the main leaf of the car spring, broken into three pieces, together. Last, they covered this with the goatskin, which, as it dried and shrank, hardened into a rawhide sleeve. Approximately four hours later they were back on the road (L.S.B. Leakey, 1974, p. 250).

The following year, in 1952, Louis and Mary purchased five acres in an area known as Langata, at the time in the countryside, about twelve miles outside Nairobi. They built a larger home with rooms surrounding a courtyard that provided security and allowed the dogs to run freely. The area had a resident giraffe and lions were frequent visitors. Other wild animals could be seen in the distance and, to the west, the sun set over the Ngong Hills.

## POLITICAL TURMOIL

In spite of their new home, the 1950s were difficult years. Kenya, still British East Africa and a colony of Great Britain, was in political turmoil. As early as 1949 Louis, along with others, tried to warn the government about a rising rebellion, but the British colonial governor was cavalier about its danger. In a 1966 reprinting of his 1936 book, *Kenya: Contrasts and Problems*, Louis added a prologue that said it was "the general growing feeling of fear, distrust and hatred, coupled with the complete failure of many officials to understand what was happening that led to the development of what became known as 'Mau Mau'" (L.S.B. Leakey, 1966a, p. vii). Eventually Mau Mau forced the British government to declare a state of emergency from 1952 til 1960.

Mau Mau was a secret organization that was built on belief in magic and ritual. To make identification difficult, the group did not name itself publicly or in written documents and there are still only guesses about the origin of the name. As Mau Mau support grew, so did confrontation with the colonial government. British arrests of suspected leaders worsened the situation. One of those arrested was Jomo Kenyatta.

Jomo Kenyatta had returned to Kenya after studying abroad. Widely admired by indigenous Kenyans and a Kikuyu Louis had known for many years, he was not considered one of the radical members of the movement for independence but neither did he renounce Mau Mau. Eventually even Kenyatta's personal popularity probably could not have stopped the uprising. Warfare, with violent, brutal killings erupted and Mau Mau held their own with homemade and captured weapons against well-trained and equipped British troops. The indigenous Kenyans suffered most, but those in the white community, whose deaths numbered under one hundred, were terrified. Mau Mau proved that indigenous Kenyans would fight and die for their rights. Ultimately, Mau Mau led the way for Kenyan independence, but the country was in turmoil for much of the decade.

During the rebellion, Louis was in a precarious situation emotionally and physically: he loved and identified with the Kikuyu people, but as curator of the museum, he served as an employee of the British government. The Mau Mau put a price on Louis's head, and he and Mary carried a pistol or revolver with them at all times and stationed trusted guards to protect the children and their new Langata home. Louis and Mary and their immediate family were not harmed, but the rebellion directly touched their lives when an elderly cousin and his wife were gruesomely murdered. Louis also served as an interpreter for the trial of Jomo Kenyatta after his 1952 arrest. Interpreting was a no-win situation because both sides considered Louis biased in a trial that many thought was rigged. Kenyatta was sentenced to seven years at hard labor. Eventually the rebellion died down but only with Kenyatta's release in 1961 was Kenya able to move beyond the bloodshed.

Mary, looking back thirty years later, well summarized the Mau Mau movement when she wrote that it was the "violent side of the Kikuyu's people's legitimate desire for independence and of their grievances, mostly real, about the European seizure, domination and exploitation of land that should have been theirs" (1984, p. 111).

## RETURN TO THE PAST

In the midst of this political turmoil, Mary and Louis were anxious to begin more intensive explorations at Olduvai and the remote camp in Tanzania offered safety from the Mau Mau threat. After digging a trial trench, in 1952 they began detailed excavation that continued for thirty years.

Even with increased funding from Boise and from the media hype generated by *Proconsul*, their finances barely improved. Publication of Louis's 1951 book *Olduvai Gorge* had been delayed and it was soon outdated by new information, a difficult pill to swallow since part of their income depended on the sale of books. Between 1951 and 1958 Mary and Louis focused their efforts on Bed II, the second oldest of the five major sites they identified at Olduvai, but their limited resources prohibited a thorough job. The financial situation was complicated by Louis's being, as Mary described him, "one of nature's financial innocents" (1984, p. 157).

Olduvai was rich with tools and bones of animals and they felt positive they would locate a mother lode of human fossils. In 1952 they found such a great cache of tools and fossils of huge extinct beasts that Louis called the area where they dug the "Slaughter House" (Morell, 1985, pp. 176–77). Yet the years from 1952 to 1959 were frustrating. Louis and Mary underestimated the difficulties they would encounter at Olduvai, an area that stretches across vast miles of land. Concern always loomed about repeating the disappointment of Olorgesailie, where they found a wealth of evidence to human presence but no fossil proof of early humans. There were other problems in working in a remote, harsh location. Water had to be transported from nearly thirty miles away. One afternoon in 1953, Louis suffered heat stroke and overnight his hair literally turned white and his appearance aged noticeably.

## PERSONAL TURMOIL

Amidst all of this, the Leakeys' marriage began to disintegrate. Biographers have generally blamed the crumbling of their marriage on Mary's later career success at Olduvai, Louis's traveling to publicize their finds and garner more funding, and financial pressures when Louis put their limited resources into other projects. Mary herself gave 1968 as the year their relationship began to fray (1984, p. 140), but in reality the marriage suffered a number of severe jolts long before that. Louis was charming and charismatic and enjoyed the company of women. He was also extraverted, impetuous, brilliant, and restless, with a huge ego. He was attractive to many women and often traveled alone from the time of his days in intelligence during World War II. Gossip that he acted on his inclinations may have been true even then.

Louis's most serious liaison was with a young British woman named Rosalie Osborn whom he met in England. When she moved to Kenya in 1955, Louis hired Rosalie as his secretary. Mary and Louis fought furiously over his affair with her. Mary realized this relationship with Rosalie was a threat to their marriage and began drinking heavily. Louis, angered by Mary's behavior, sought continued solace not with drink but with Rosalie. The possibility of divorce was real and the children were aware of their parents' arguments. Although Jonathan was able to mentally withdraw from the chaos in the house and Philip too young to comprehend, Richard, at age eleven, was affected and begged his father not to go.

Richard's love of horses and willingness to risk-take probably helped keep his parents together. A galloping horse threw Richard and he suffered a skull fracture. During the weeks he lay in bed, with the outcome uncertain, Louis realized that the arguing and shouting hurt Richard's chances of recovery and decided to remain married to Mary.[2]

## INTRODUCING RICHARD

The year 1955 set the course of Richard's life in a different way. He entered a new school that was modeled in the Victorian British tradition, with a focus on strong discipline and maintaining a stiff upper lip, regardless. Richard's response to such a disciplined regime was exactly the same as his mother's—he hated the academic world from then on. The very first day, to gain favor with the older boys, a boy from his previous school pointed him out as someone who liked blacks, not a popular position to take at that time of racial segregation and Mau Mau insurgency. He was tossed into a small, locked wire cage, and treated like an animal, poked with sticks and even urinated on. The boys eventually went off to class and when he was eventually found and rescued, he was blamed.

Other tricks soon followed. He might have redeemed himself if he had been athletic, but he was not good at team sports. The only time he excelled was in a rugby game when he accidentally was given the ball and, to save himself from being tackled, ran hard, luckily in the correct direction, and made the only score of the game.

Like his mother, Richard survived by creatively outwitting the school establishment. He and other seniors built a subterranean room in which four of them at any one time could smoke and enjoy a little home brew. They emerged one day to discover one of the masters at the entrance and soon their underground hideaway was packed with stones and filled with water.

After Louis broke off his relationship with Rosalie and while their three sons were maturing during the late 1950s, the family enjoyed some of their happiest times. All of the boys were academically uninvolved, but Mary noted in her memoirs that she was hardly one to criticize them on that score. Richard's interest in horses led the entire family to become absorbed with horseback riding. They formed a pony club with all the children in the Langata neighborhood. In addition to horse shows, chasing zebras—easy to lasso—and giraffes—impossible to lasso—was fun. They always released the animals unharmed; the riders were the ones in possible danger. Going a little further from home offered the chance to chase rhinos. Explaining the thrill, Richard wrote, "Chasing rhinos was the ritual proof of teenage skill, and bravado. . . . Our aim was to get alongside a fleeing rhino, give the thundering beast a good slap on the rump which usually annoyed it so much that it turned and we would then have the excitement of being chased ourselves" (1986, p. 41).

Without electronic games or even television, the evenings at the

Olduvai camp were passed with games like chasing spring-hares, which are not rabbits but small nocturnal creatures that hop like a tiny kangaroo. Driving slowly across the Serengeti, when a hare was spotted in the Land Rover's headlights, they would chase it on foot. Anyone who caught one of the little animals would hold it for a minute before releasing it. Then the hare hopped off and the game began anew. Of course, if the driver got involved in the action and charged after the hare, the driverless vehicle went sputtering off until someone caught up with it or it stalled.

Louis also had contact with the children from his first marriage. When Frida and Louis divorced, Frida did not ask for money for herself until Louis's financial situation improved, which of course was not ever. She didn't remarry and elected to rear Colin and Priscilla alone quietly in England. At age nineteen when Colin met his father for only the second time since he was a baby, Louis took Rosalie along to the pub where they met, a situation that, to Colin's dismay, didn't seem to bother Louis but did Rosalie and him (Morell, 1995, p. 244). Priscilla learned she had three half-brothers only when she looked her father up in *Who's Who* (Morell, 1995, p. 244). Priscilla and Colin finally got to know their half-brothers in 1963 (R. Leakey, 1986, p. 187).

## EXPANDING INTERESTS, FRIENDS, AND FAUNA

The Leakeys were working parents and Louis's scientific interests ranged far. Although he spent his entire career attempting to prove that humans originated in Africa, he saw relationships among various areas of science. He suspected that, just as humans evolved physically, their behavior also evolved and he enthusiastically set about finding the right people to investigate relatives close to *Homo sapiens*. In 1958 he established the Tigoni Primate Research Centre for the study of monkeys in captivity but also championed study of nonhuman primates in their natural habitat. Jane Goodall was the first to attract worldwide attention.

Early in 1957 twenty-three-year-old Jane Goodall traveled to Nairobi because of her lifelong interest in animals. When she made an appointment to meet with Louis, she didn't realize he had suffered through numerous failures and dead-ends in finding someone whose study of today's apes would help explain *Proconsul*'s behavior during the Miocene.

Louis piqued Jane's interest in studying chimpanzees when she spent a month at Olduvai. He described the excitement of studying a lone colony in a distant area along a lakeshore. She had no academic credentials and her gender in the mid-twentieth century was a strike against her. However, he promised he would obtain funding for several months so she could study chimp behavior. Jane by now knew she was preceded by several others who had failed and the idea that she too might let Louis down caused her difficulty sleeping, but she was thrilled at being asked and desperately wanted to go. Finally, in the summer of 1960, after interminable delays for a variety of reasons, she set off for the Gombe Stream Chimpanzee Reserve. Actually, since it would not do for an unaccompanied female to go alone, Jane, her mother, Vanne, and Dominic, a cook, set off. Louis, as usual, had wrangled funds from a unlikely source, a foundation begun by a tool-making firm in Des Plaines, Illinois, and, also as usual, had purchased equipment with saving money in mind—cheap, secondhand binoculars, a tent, and tin dishes.

Jane Goodall has made immense contributions to knowledge about humans' closest relative with whom humans share 98 percent of their DNA. In patiently living with chimps for over forty years, she has discovered they display many traits, such as manufacturing simple tools and fighting rival communities, that are comparable to those of humans. Also like humans, their toothy smile is usually friendly or can indicate nervousness or fear. Other social behaviors that are humanlike, such as expressing sympathy, indicate evolution from common origins.

Jane was the first of Louis's three protégées who became world famous for studies of primate behavior in their natural habitat. Her success paved the way for Dian Fossey and Biruté Mary Galdikas.

## NOTES

1. The chief secretary of the Kenyan government stipulated in a letter to the British Museum of Natural History that the skull belonged to Kenya and,

when an adequate display facility was available to house it, it should be returned.

When Richard Leakey became director of the National Museums of Kenya, he tried without success to have it returned. The British Museum denied that such a letter existed until 1982, when Hazel Potgieter, Mary's secretary, found a copy in the National Archives. The skull is now on display at Nairobi's National Museum.

2. As a footnote to this episode in their lives, Rosalie, who never married, upset the family a quarter of a century later when she placed a marble head-stone at Louis's gravesite after Mary and the sons procrastinated in erecting a monument. "ILYEA," the way Rosalie had signed her letters to Louis, is engraved at the bottom, code for: "I'll Love You Ever Always." The tombstone angered and offended family members but removing it seemed "far more scan-dalous" (Morell, 1995, 404–405).

# CHAPTER 6

# HUMAN HISTORY CARVED IN BONE

## *(1959–1971)*

Geologist and paleontologist Jon Kalb envisioned fame and fortune when he found stone tools and what he hoped were fossils from early humans who made the tools. Louis and Mary agreed to look at his collection.

I laid the fossils and artifacts out on a coffee table between us.

"Aha!" said Louis.

My pulse quickened.

He picked up the fossils one by one, carefully turning each over and over.

My excitement was at fever pitch. . . .

Finally, Louis said, "Fossil pig. Yep, fossil pig. Yep, I'm sure they're pig."

Crestfallen, I picked up one of the artifacts from the table and handed it to him, and said, "Well, is there any chance the *pig* could have made the stone tools?"

Louis got a good laugh out of that, but Mary, an archeologist, was not amused. . . .

—Jon Kalb, *Adventures in the Bone Trade*, p. 46

## "WHAT BIG TEETH YOU HAVE!"

Although the state of emergency in Kenya was not lifted until 1960, the summer of 1959 found the trauma of Mau Mau receding into the background and Kenya moving toward independence from Great Britain.

Perhaps, Louis and Mary thought, a nearby site at Laetoli would prove fruitful. A spring 1959 trip there used most of their funds and produced only disappointment so they decided to briefly return to Olduvai. Then, in July 1959, what Louis—and many colleagues—often referred to as "Leakey luck" changed their lives.

Heselon Mukiri, Louis's trusted fossil hunter, found a hominin tooth and part of a jaw imbedded in Olduvai's limestone. Friends who produced wildlife films wanted to photograph excavation from start to finish, so the Leakeys delayed further digging for a day or two until the cameraman, a friend, arrived with his wife and daughter. Son Richard Leakey, now fourteen years old, was also arriving with them.

July 17 found Louis not feeling well. To pass time while awaiting the visitors' arrival, Mary decided to survey a site where recent rains had uncovered new fossils, far from where they would dig during filming. She was thrilled to spot several very large teeth still imbedded in what appeared to be an upper jaw of an *Australopithecus* and raced back to camp. Louis firmly believed that today's humans are not directly related to australopithecines and at first was slightly disappointed as he hoped for an early *Homo*, the maker of the tools they found in abundance wherever they dug.

Louis's disappointment soon turned to glee. There, *in situ*, waiting for its many small pieces to be found and pieced together so it could be photographed and shown to the world, was the skull, bony crest down its middle, of a creature who lived about 1,750,000 years ago. He or she who finds gets to name, based on the International Code of Zoological Nomenclature, so Louis dubbed him Zinj, an ancient Arabic word for East Africa and short for *Zinjanthropus boisei*, or "East African Man." The species name was in honor of their longtime benefactor, Charles Boise. Colleague Phillip Tobias nicknamed him "Nutcracker Man" because of gigantic teeth, but little wonder the Leakeys themselves referred to him as "Dear Boy." If *Proconsul* opened a path for the Leakeys' future work, then Zinj exposed a superhighway. His discovery, helped by his re-entry into the daylight being captured on film, led to a fantastic amount of funding in return for exclusive American publishing rights from the National Geographic Society, a collaboration that still continues with the Leakey family. Zinj put the Leakey name in lights on the billboard of the world stage, and as Louis's granddaughter Louise now says, so far as the origin of humans, the 1959 find of Zinj "really put Africa on the map." It also made stars of Louis and Mary. Donald Johanson later wrote, "The Zinj find of 1959 not only made Louis Leakey famous; it also made paleoanthropology fashionable. . . . Human fossils work a special magic . . . Fossil hominids have always had more clout than fossil clams" (Johanson & Edey, 1981, pp. 97, 98).

Someone who names new species based on small differences found among fossil specimens is termed a "splitter," while a "lumper" looks at the same fossils and uses similarities to group fossils into a few species. Louis caused amusement among colleagues when he acknowledged that *Zinj* resembled australopithecines but said he saw enough differences to give the skull a new name, even though "I am not in favour of creating too many generic names among the Hominidae . . ." (as cited in Lewin, 1997, p. 140). Louis was keenly aware that naming a new species is more glamorous and glorious and financially rewarding than simply discovering a new specimen of an existing species and almost always found a reason to give his finds a new name. Louis "of all people," Roger Lewin wrote, "was a supersplitter" (Lewin, 1997, p. 140).

The difference between splitters and lumpers is not a minor one; many of the arguments were, and continue to be, based on a paleoanthropologist's philosophy about naming, what they observe when they analyze a fossil, and ultimately how they develop a hypothesis to explain their observations. These hypotheses lead to an understanding of meaningful relationships among fossils and ultimately to how they understand the evolution of early humans. (More in chapter 9 on the present disagreement between splitter Meave Leakey and lumper Tim White about *Kenyanthropus playtops.*)

"The only way a name becomes accepted is by consensus, and there is often very little of that" (Bryson, 2003, p. 440). Louis had named the fossil officially catalogued as OH 5 (Olduvai hominid #5) *Zinjan-*

*thropus boisei.* Many scientists disagreed, and the name became *Paranthropus boisei* after the similar South African species *Paranthropus robustus.* Then, as Tattersall and Schwartz explain in *Extinct Humans*, ". . . in 1967, the taxonomic tide had turned" and Zinj became *Australopithecus boisei* (2001, p. 76). When the tide turned again, *Paranthropus boisei* returned, although some scientists still question that designation. At any rate, say "Zinj" and an anthropologist knows exactly which skull.

Ironically, Mary discovered Zinj barely one hundred yards from where Louis had led his touring scientists in 1947 during the Pan-African Congress, in the site Louis had long ago named FLK, an abbreviation for "Frida Leakey korongo," the woman she displaced as his wife. (The last word is, according to the publisher's prologue of *By the Evidence*, p. 13, "an African word for 'gully.'")

## JOINT ENDEAVORS

Because the skull was found near stone tools, Louis immediately put Zinj on the line leading to modern humans. Many of his colleagues doubted that Zinj was capable of making tools, but no matter. Within months Zinj was demoted to an australopithecine "near-man" and not the ancestor to "true man," as Louis used the terms. The person responsible for moving Zinj to a lesser role in human prehistory was Jonathan Leakey, then twenty years old. While in the field with Mary in 1960, he casually showed his mother some bones. Mary dashed to see her son's discovery and subsequently the area of excavation was called "Jonny's site." Canadian archeologist Maxine Kleindienst remembered looking at the bones that day with Mary and "Mary fully appreciated the controversies that would ensue" (Hunsinger, 1997, p. 3). They never found a complete skeleton of what appeared to be a twelve-year-old child, as Louis hoped, but Mary was especially pleased with hand and foot bones, important for assessing movement, that she and Jonathan later found from the child and an adult.

Anticipating the cry from the scientific community, Louis uncharacteristically didn't publicize their latest find until 1964, when he and Phillip Tobias, joined by John Napier, an anatomist who specialized in

Louis and Mary Leakey, 1965. *Courtesy of Bob Campbell.*

Louis Leakey discusses a fossil specimen with his son Richard. *Photofest*.

A portrait of Richard and Meave Leakey at a table with some primate fossils at the Nairobi National Museum in 1986. *Courtesy of Bob Campbell*.

Louise Leakey with her mother, Meave Leakey, out in the field at Ka-
napoi—the area where the fossil hominid mandible *Australopithicus ana-
mensis* was found in 1994. *Courtesy of Bob Campbell.*

Louise excavating fossil hippo skull at East Turkana.
*Courtesy of Scott Bjelland.*

hands and feet, named it *Homo habilis*, or "Handy Man." In doing so, they upset the existing view of the genesis of humans almost as much as paleoanthropologists upset the biblical view of creation. Louise Leakey calls *Homo habilis* "one of Louis's most controversial specimens" and "one of the questions that needs to be resolved" (personal communication, October 10, 2003). Two issues have invited questions and contention for half a century:

> Could several species of early humans have existed at the same time?
>
> Should *Homo* be defined in terms of anatomy or should the definition include other traits and skills?

Because *H. habilis* had a larger brain than australopithecines and a skull more like modern humans, Louis concluded that Handy Man was of the human lineage. But the bones were found at a lower level than Zinj, indicating an australopithecine and an early human lived in close proximity to each other prior to two million years ago, give or take a quarter million years, which, evolution-wise, is at about the same time.

Two species of early humans living as contemporaries? Not possible, many scientists of the time chimed in unison. The single-species hypothesis of evolution dominated the world of paleoanthropology in the 1960s and still has its advocates. Since early humans, this theory hypothesizes, were unique in skills like walking upright and using their brains to make tools, only one true ancestor could have existed at any one time, with the family tree having only one living branch going upward from the break with the great apes to today's humans. Other genera are represented by many species; pigs, for example, are an ancient species with lots of diversity in the fossil record and so have a bushy family tree. Louis believed, because multiple species of other animals exist at the same time, that different species of humans could logically live at the same time. He felt the belief that humans alone are the only species to exist in isolation was based on a religious attitude that humans are a special species. At any rate, most anthropologists had to take a big gulp to digest the idea that two species of early humans lived contemporaneously.

Louis and his colleagues also caused a furor because, in assigning

*H. habilis* to the genus *Homo*, they toyed with the definition of *Homo*. Naming a new species is sure to trigger arguments; changing the definition of a genus is certain to start a war. Today context—cultural issues like use of tools, language, and ability to control fire, and whether the specimen lived in rain forest, savanna, or desert—are considered relevant, but fossils are still given names based solely on morphology; that is, form and structure. Morphology presumably reflects the species's place within an evolving lineage. Description half a century ago relied only on anatomy until *H. habilis* was named: "The diagnosis offered by Leakey, Tobias, and Napier was more comprehensive, and included habitual bipedal posture and gait, a precision grip, and a brain capacity much smaller than previously proposed" (Lewin, 1997, p. 146). Although they didn't specify the use of tools, the name they selected, *Homo habilis*, or Handy Man, suggests tool use. And the accepted size of brain to qualify as *Homo* was reduced; *Homo habilis* became a small-brained toolmaker.

Skewering colleagues with punctuation marks is not unknown in paleoanthropology and Le Gros Clark, usually a steadfast supporter of Louis, infuriated him by writing "Homo habilis" with quotation marks to make it seem like a nickname, instead of the conventional *Homo habilis*. *Homo habilis* is now generally regarded as a legitimate species but isn't universally considered a direct ancestor of humans. Interestingly, *Homo habilis* became more accepted after Richard Leakey found cranium KNM-ER 1470 at Koobi Fora in Kenya in 1972. Except that 1470 is now designated *Homo rudolfensis* by some anthropologists, *Kenyanthropus rudolfensis* by others and *Australopithecus rudolfensis* by still others. Richard Leakey has said, "*Homo habilis* is such a grab-bag mix of fossils; almost anything around two million years that doesn't fit the robust definition has been tossed into it" (as cited in Morell, 1995, p. 549). Jeffrey McKee wrote, "Most scholars in this peculiar branch of anthropology now accept the legitimacy of *Homo habilis*, but almost all disagree on which fossils represent the species" (McKee, 2000, pp. 79–80). He calls the naming and reorganizing of fossils "taxonomic hell" (p. 79). Then in a speech commemorating Louis's 100th birthday, Meave Leakey said she didn't even want to mention *Homo habilis* because it would present a distraction. Undoubtedly wise!

## POLITICAL PEACE

The personal appeal of Jomo Kenyatta, who provided inspiration for rebellion against the British government in the 1950s, grew during his years in prison. It was clear after his 1961 release that he would become Kenya's first president when the colonial period ended on December 12, 1963. Whites remaining in Kenya, including Louis and his family, had no idea how they would be treated when Kenya gained independence and they went from a position of power to become a small minority group. Most white Kenyans who elected to remain in the country must have been greatly relieved when Kenyatta offered an olive branch. Kenyatta's philosophy was a healing balm for the country and laid the supports for a democratic republic that has fared better socially and economically than many African nations.

Richard Leakey pointed out in *One Life* that one of the few positive results of colonialism was the government's demand that the Swahili language be the *lingua franca* of the multiple tribes and other people living in the country. "The welding of a united Kenya after Independence from such diverse indigenous elements would have been considerably more difficult had there not been a common African language with which to communicate" (p. 14). As children, indigenous Kenyans first learn their tribal tongue, then the Swahili language, and then English. Most children with roots in the Western world learn basic Swahili in addition to English. Today travelers to any part of Kenya are welcomed with a smile and "Jambo!"

## LIFE GOES ON

By the early 1960s the three Leakey sons were independent young adults. Philip was still in school, although he soon went to secondary school in England and, when not yet sixteen, left after the first semester. By 1963 Jonathan was already on his way to a career as a herpetologist when he married Mollie Knights-Rayson.

In 1961 Louis gave Richard the choice of attending school or supporting himself. Richard, not knowing what to do with his life, chose the latter. For a while he trapped and sold wild animals, including baboons. He also collected carcasses of dead animals whose bones he then boiled, bleached, and labeled, for sale to universities and museums. He helped manage the camp at Olduvai and a year later started his own safari company, leading people eager to see the African bush with Louis Leakey's son as their guide.

## MOVING APART

*Proconsul* was responsible for immediately bringing Louis and Mary closer together, but they moved further apart after the discovery of Zinj. More and more Louis traveled to publicize their work, more and more he poured limited funds into a variety of undertakings, and more and more he sought females to provide affection and boost his self-esteem.

"Speaking before a packed lecture hall in his staccato-like voice, punctuated by rapid inhales, he cast a spell, making each listener believe he was speaking only to him or her," was how American paleoanthropologist Donald Johanson described Louis's lecturing (1999, p. 122). Dr. Leigh Van Valen, Professor in the Department of Ecology and Evolution at the University of Chicago, remembers one of his many speeches. In speaking at Columbia University in the mid-1960s, Ralph Holloway, called "probably the world's expert on modern and fossil hominid brains" by Roger Lewin (1997, p. 67), gave Louis a flowery introduction, Van Valen remembers, "climaxed (for me) by saying, 'Here is the man who has made Leakey a household word!' Quite deadpan—I don't think he realized how it sounded, and no obvious realization by the audience except for a couple of us students" (personal communication, February 7, 2004).

Regardless of how he was introduced, Louis masterfully carried his message to a public eager to hear the latest information from the field about humankind's beginnings. He also used newspapers so frequently to publicize Leakey fossil finds that he was often called the "Abominable Showman" and *Punch* magazine referred to their camp as "Oboyoboi Gorge" (Hellman, 1998, p. 164).

In addition to a hectic schedule of trips to the United States in which he made multiple speeches and scurried to find funding, Louis spread himself thin by investing time and money in other projects. He resigned in 1961 after twenty years as curator of Nairobi's National Museum, but he juggled many other balls. He established the Centre for Prehistory and Palaeontology and was involved with supervising the Tigoni Primate Centre, writing articles, serving as a trustee for the East Africa Wildlife Society, and overseeing numerous and diverse projects. He started a Society for the Prevention of Cruelty to Animals in Ethiopia. He made visits to Israel to check progress on excavating artifacts from a Paleolithic site at Ubeidiya. He even studied the medical properties of zebra fat, which some tribes use to treat tuberculosis. In addition to Jane Goodall's Gombe Stream Research Center, he started and supported primate projects by Dian Fossey and Biruté Mary Galdikas.

Dian Fossey was an Occupational Therapist when she approached Louis in 1966 about studying gorillas. She was amazed that he remembered her from a safari she had taken three years before. Although perhaps *forgetting* he had met her in 1963 would be more surprising; at Olduvai she not only sprained her ankle, but she vomited on some fossils. Later on that same trip, despite the sprained ankle, she climbed to the mountain home of gorillas, where she spotted six of them.

At her second meeting with Louis, Dian convinced him that she was plucky and resolute about studying these huge relatives of humans and Louis both found funding sources for her research and provided an entrée at Cambridge for her work on a doctorate. Dian arrived at her camp in the Congo the first week in January 1967. In Louis's first letter to her he warned that he was worried about danger from a poacher. His words were eerily prophetic; she was murdered December 26, 1985, in her cabin in Rwanda where she had moved. Her murderer has never been found, but someone angry about her determined work to stop the wanton killing of gorillas is logically a possibility and graduate students and coworkers have also been suspects.

Biruté Mary Galdikas is not so well known as Jane Goodall and Dian Fossey, partly, Galdikas said in a 100th birthday tribute to Louis, because her name is so hard to pronounce. Galdikas became interested in monkeys at age six, when she read a book featuring the quirky monkey *Curious George*. She was encouraged to contact Louis in 1969, after she heard him lecture about Fossey's work. Although Louis was initially discouraging when Galdikas told him she wanted to study orangutans, he was convinced and again managed to cobble together several small grants. Unlike many of his protégés who lacked credentials, she had a master's degree in anthropology. Now Dr. Galdikas, she has studied orangutans in their natural environment for over thirty years and, because poaching and destruction of their habitat have placed orangutans at risk for extinction, continues her work at Camp Leakey in Indonesia.

While Louis's fund-raising efforts provided the capital and his various expenditures used up a lot of it, Mary kept the coffers filled with fossils. She remained at Olduvai most of the 1960s to dig and write. Even in the evenings, when others in the camp were relaxing, she preferred her solitary work of studying and categorizing fossils. Mary was the meticulous worker and Olduvai was not just her camp but her home.

Although Mary and Louis remained married, their relationship became primarily professional. A cartoon by Gary Larson shows a paleoanthropologist, right arm embracing a woman and left hand holding a skull, with the caption "The anthropologist's dream: A beautiful woman in one hand, the fossilized skull of a *Homo habilis* in the other" (Larson, 1989, p. 214). The chunky mustachioed man in baggy pants and shirt with pockets looks suspiciously like Louis. Just as Louis courted paleo-controversy his entire life, he also courted feminine companionship.

Louis fell in love with Jane Goodall and showered her with letters of love but, to his credit, when she rejected him romantically, he retreated but still mentored her and fought vigorously to obtain funding for her research. His relationship with Dian Fossey was a different matter. He expressed fervent love to Dian, with whom he shared an on-again, off-again relationship until his death. These are two better-known names among a bevy of females, mostly young women awed by Louis's fame and intrigued by the possibility of entrée into the scientific field through him. He charmed them intellectually and charmed many of them romantically.

Louis's relationships were especially galling to Mary. He purchased supplies and equipment the young women needed from personal funds. Even more annoying, he gifted them with remembrances like roses and plane tickets and cameras and he gave Dian Fossey a ruby ring and bush jacket while Mary "was living an utterly Spartan life at Olduvai" (Morell, 1995, p. 384). Mary chalked up Louis's appeal to women as mostly a matter of "chemistry" (1984, p. 42). She knew that in the 1930s, when he felt his first marriage was a mistake, he saw other young women, even though Frida was pregnant with their second child; gradually he focused only on Mary. In *Disclosing the Past*, Mary wrote honestly that sometimes she arrived at their Langata home to find him with someone. His actions hurt her but even worse was her feeling that he had a "deplorably low standard" (p. 158) in choosing women for his romantic liaisons.

Mary retaliated by turning to alcohol, a habit Louis hated and one that diminished Mary in the eyes of many colleagues, especially since, when Louis was with her, he usually deferred to Mary and quietly treated her with respect even if she belittled him. Ralph Holloway remembers Louis as "extremely kind" to him and his family. But Holloway also remembers that Mary, having drunk too much, didn't find the situation humorous "when her Dalmatian dog, about which we were talking, got up, went over to my camera bag, and pissed all over it! I laughed uproariously, she threw me out of her camp" (personal communication, September 8, 2004).

In one of Louis's letters to Dian Fossey, Louis wrote, "I love you and love you so and there are no words that can describe the peace and calm—happiness that are springing from our love of each other" (as cited in Morell, 1995, p. 330). Another young woman wrote of Louis that he "desperately needs affection" (as cited in Morell, 1995, p. 329n). There were complex reasons for their alienation, but peace and calm, happiness, love, and affection were not what either Mary or Louis now found in their relationship.

Louis's determination to support fossil expeditions in the Calico Mountains, near Barstow, California, was a final breach in his relationship with Mary. Money was always an issue, but when he used the money for what became known as the Calico project, Mary considered it an utter waste. His deteriorating physical condition and his adoration from Americans seemed to play into his enthusiasm for an investigation that dragged out more than six years. He was determined to convince the world that stones collected from the site were tools made by early humans who lived in North America 80,000 years BCE, a rather outrageous claim then and it still is today. He organized an international conference at Calico in October 1970 to show and discuss artifacts found there. Mary was not invited nor would she have gone. Participants, many of them both friends and colleagues, did not dispute him. Not because they believed the artifacts proved humans inhabited North America that long ago, but because they sensed he no longer was capable of rational scientific reasoning and didn't want to hurt him. Excavations are still going on at the Calico site, the only archaeology project in North America in which Louis was involved.

## RICHARD ON THE MOVE

During the years 1966 to 1970, Richard married (1966), found the fossil-rich area of eastern Lake Turkana (1967), gained funding for an expedition of his own (January 1968), acquired his first administrative position at the National Museum as assistant director (May 1968), learned he had a life-threatening illness (May 1968), began expeditions at Koobi Fora (summer 1968), became a father (spring 1969), divorced (1969), established the Wildlife Clubs of Kenya Association for young people (1969), took over the museum's directorship (1970), acquired a pilot's license he had let expire (1970), remarried (1970), and established a permanent camp at Koobi Fora.

Richard met Margaret Cropper, his first wife, at Olduvai in 1961 when she worked closely with Mary, and he followed her to Scotland, returning to school when she did. Richard fared well with his academics but his safari business somewhat soured and spending a number of years in a classroom didn't seem terribly important. In fact, while facing the challenge of transporting a fossil elephant carcass from Baringo to Nairobi—no small job—he forgot about the university's application date.

Richard and Margaret married in 1966 and she went with him on the first expedition to Lake Turkana in 1968, but their marital troubles were soon obvious to their friends. Margaret disliked his using the family name and contacts for career advancement, a common practice in Kenya, instead of earning an academic degree. He disliked formal education and saw networking as an intelligent option. She criticized him in public, to which he might retort but then, not wanting to sound like his parents, hold comments inside. In 1969, Margaret and Richard separated and divorced. Although Richard's actions mirrored his father's in leaving a pregnant wife and divorcing when the child was a baby, his motivation was different. Years later Richard wrote in *One Life* that he divorced Margaret because he didn't want his daughter Anna, born in 1969, to "have the ghastly experience of living with fighting parents" (p. 111).

Meave Gillian Epps wasn't the cause of Richard's divorce, but she was a catalyst. Meave initially made her start by working at Louis's Tigoni Primate Research Center, where Louis suggested she might find a dissertation topic of interest. She selected the limb bones of colobus monkeys and received her doctorate in 1968. The topic was a fortuitous one, because Richard hired her to help him write an academic paper on an extinct giant colobus monkey and soon their relationship was personal as well as professional. In fact, partly to stake out his own territory and avoid competition with his parents, Richard anticipated spending his entire career studying fossil monkeys, but, just as Johanson said that human fossils get more acclaim that those of clams, Richard soon realized that little fame goes to finders of fossil monkey bones.

In the summer of 1967, Richard, working in the Omo valley of Ethiopia as part of an international expedition, found hominin fossils, partial skulls called Omo I and Omo II, but he estimated they were only 130,000 years old. Ironically, by 2005 geological research in Ethiopia and better dating techniques supported an age of 195,000 years for the skulls. Omo I has a modern look and Omo II, a skullcap, appears more primitive, but both are thought to have existed at the same time. These fossils immediately became the oldest specimens of *Homo sapiens* yet found and, because the age agrees with genetic analyses of population, give weight to the argument that Africa is the birthplace of modern humans. In 1967, however, these partial skulls gave Richard little reason to become excited.

That summer Richard's excitement came from another type of find. He had returned to Nairobi to attend to safari business concerns and purchase supplies. On the flight back the pilot diverted the plane to avoid a thunderstorm and flew low over the eastern shore of Lake

Turkana, then known as Lake Rudolf. Richard's keen eyes spotted what appeared to be fossil-rich sediments. The next day he hired a helicopter and decided the area looked promising enough from the air to land. When he immediately found some fossils, Richard knew he would return the next year. He was so enthusiastic that he later realized he hadn't identified the exact location and wasn't even certain if he found the fossils in Ethiopia or Kenya! But by spring 1968 Richard had surveyed, with care this time, the northeast shore of Lake Turkana, obtained funding from the National Geographic Society—with his proposal accepted over one from his father—and was leading his own expedition based at Koobi Fora, along the lake's shoreline.

That same spring of 1968, despite being in his early twenties and a school dropout, Richard campaigned for an administrative position at the National Museum of Kenya, and, with help from a powerful friend and the Leakey family name, he was appointed. Even before taking the job offered, with characteristic boldness and self-confidence and help from a number of loyal Kenyan museum board members, Richard parlayed the part-time job into a position as the museum's administrative director.

Although self-confidence has always been his strong suit, chutzpah couldn't help Richard's health. Three days before beginning the job, he developed a sore throat. The doctor who examined him diagnosed his difficulty as kidney disease set off by a throat infection and prescribed six weeks of bed rest to avoid permanent damage to his kidneys. Richard told only Margaret and, undeterred by either the doctor's diagnosis or prognosis, continued a busy schedule. Then, while on a March 1969 trip to the United States, he visited a specialist whose opinion was even gloomier. Because of damage to his kidneys, he would suffer kidney failure within the next six months to ten years and would immediately need an artificial kidney or a kidney transplant.

While on the trip, Richard also worked to find a geologist with the ability to date the sandy sediment along the eastern shore of Lake Turkana. The young woman who accepted the mission of working in this difficult area of northern Kenya was a Harvard doctoral student named Kay Behrensmeyer, now research curator in the Smithsonian's Department of Paleobiology, and she quickly became an asset to Richard's team. He also gave his first public lecture to an audience of about three thousand at the National Geographic Society and realized that, like his father, he very much enjoyed public speaking.

Back in Kenya, Richard kept his illness a secret and threw himself into his work at the museum and fieldwork at Koobi Fora, where he immediately found a trove of fossils. Why was fossil hunting so lucrative there?

All the adjectives commonly used to describe a desert apply to Koobi Fora—scorching hot, stark, barren, forbidding, bleak, arid. This landscape is pierced by Lake Turkana, the world's largest permanent desert lake, whose water shimmers with shades of jade. About 1.8 million years ago, however, the banks of a river that flowed gently through Koobi Fora were dotted with trees. Giant turtles, crocs, and hippos wallowed in the river's water and animals of all types roamed the swampy grasslands. Short-necked giraffes, huge baboons, pigs, rhinos, horses, monkeys, and elephants made their homes in the nearby forests, vigilant of predators like the saber toothed tiger, with its long, curving teeth. A volcano overlooking the area burped ash that settled on the earth's green floor or became part of the riverbed, belches that today are helpful in obtaining accurate dating of the soil in which fossils are found. Fossil hunting at Koobi Fora has been as profitable as Richard imagined when he first viewed the area from the air. During the first field season at Koobi Fora, Richard and his team, including Meave, systematically surveyed what he called an "incredible area" and Richard wrote, "One of our biggest problems was the sheer quantity of fossils!" (R. Leakey, 1986, p. 98).

Richard envisioned expeditions by camel, an idea that sounded better than it proved, although the camels could travel over desert soil without damaging exposed fossils, as Land Rovers did. However, camels are stubborn and slow, unless they're running away, and, despite the Lawrence of Arabia aura Richard presented, his burnoose flapping in the wind as he led his team onward, the camels were a temporary innovation. Both game and gangs eyed the camels. Lions saw camels resting at night and thought dinner; gangs of bandits saw the small group astride the awkward animals and thought money. On a trip to get water, about eighteen men wearing little except loincloths but carrying lots of weapons and ammunition confronted Richard. They warily watched the lonely rider as he lifted his own rifle overhead and then filled his water containers. As he headed back to camp, they melded again into the desert.

Richard and Meave did make an important find while aboard their camels. One blistering day in 1969 when Richard was eager to return to camp and a cool drink, he spied a complete cranium of *Australopithecus boisei*. Mary was in camp when he brought it back and she proclaimed it magnificent, a thrill almost as great as when he first saw it peering at him from an amazing resting place atop the soil.

The director of the National Museum left in 1970 and, when no replacement was named, Richard quietly gained control. He increased the scope and outreach of programs for the museum through a five-year program that moved all prehistory sites, with their budgets, under his jurisdiction. He acquired regional museums. He established wildlife clubs in the schools to raise interest and awareness about wildlife conservation and increased the budget by a huge amount. A policy was put in place to put indigenous Kenyans in positions of responsibility. By 1970 the ambitious second son was poised to make his mark.

## TROUBLES WITH FATHER

In Richard's work to strengthen the museum, coming into conflict with Louis was inevitable, partly because of his father's advancing age and partly because of Richard's actions.

The years of ignoring his body were catching up with Louis. He was overweight, had lost many of his teeth, and walked with a cane because of a painful right leg. Then, on January 31, 1970, while at Olduvai, Louis's health took a serious turn. He refused to admit he was having a heart attack. Instead, he took a plane to London, where Jane Goodall's mother, Vanne, at whose flat he always stayed while there, immediately contacted a doctor. While in the hospital he suffered a more severe heart attack and could not return to Kenya until July.

The next January Louis suffered an even more painful attack, not from his heart but from bees. While visiting ruins of a historic site he was attacked by thousands of bees and in the horror of the incident he fell, hit his head, reinjured his leg, and suffered a temporary loss of vision. Louis later said he stopped after counting eight hundred bee stings. He was near death but recuperated and reported to Vanne Goodall, "I must be very, very tough" (as cited in Morell, 1995, p. 371). He gave no intention of slowing down and was soon making plans for yet another lecture tour in the United States.

Louis had tended to be anxious and suspicious before the heart attack; afterward he seemed almost paranoid. He believed Richard was undercutting him, which was to some extent true. At times Richard certainly didn't consider his father's feelings. In 1968 when Richard started a program to make casts of fossil specimens in Nairobi with Kenyan trained staff, he eliminated the British company with whom Louis had contracted. This company was, as Richard knew, operated by Jane Goodall's sister, and the contract helped the Goodall family, which had been so generous in helping Louis. Richard later realized, "I was doing the right thing but largely for the wrong reasons; in other words, I was acting at least partly out of spite even though the programme I initiated did benefit the Museum and Kenya" (R. Leakey, 1986, p. 105).

One of the sites Richard placed under his museum umbrella was the Centre for Prehistory and Palaeontology, his father's "baby" and previously an autonomous institution. These situations and similar ones caused family arguments. In addition to Louis's physical condition, Richard's kidneys were now at war with the rest of his body. He ached all over, his rising blood pressure in turn caused migraine headaches, and he easily became irritated and testy, but his father had no knowledge of Richard's health problems. In summarizing the relationship of Louis and Richard during the late 1960s and early 1970s, Morell wrote, "The ongoing quarrels between father and son were now legendary, and for every fossil discovery that drew them together, some new administrative issue cast them further apart" (1995, p. 388).

# CHAPTER 7

# SKULL DRUDGERY AND SKULLDUGGERY

## (1972–1977)

> It's increasingly important to know that we as humans have a
> common ancestor and it's even more important to know that we
> have an African origin.
>
> —Louise Leakey, television interview; Eddings, 2003

## LOUIS AND LOUISE

The year 1972 brought three major changes in the Leakey family. Meave, who married Richard a year after his divorce from Margaret was final, gave birth to their first child, Louise, on March 21, 1972. When her parents were at Koobi Fora she slept in a netting bed that swung from the rafters to create a breeze and sometimes sat in a pan of water to cool her.

In July Bernard Ngeneo of Richard's team discovered the skull designated KNM-ER 1470, meaning the hominin fossils found at East Rudolf by a member of the Kenya National Museum team numbered 1,470. Even if 1470's name was unexciting, its date was. The soil in which it was found was below sediment dated at over two million years, much older than anyone thought possible. Sieving for the many shards took weeks and it took more weeks for them to be pieced together. Like his father, Richard was blessed with a wife skilled in reconstructing skulls. He credited Meave with reconstructing it so the pieces, after wearying days of work, fit perfectly, without missing fragments that would make the large brain size questionable. Over thirty years later, at the 2003 Louis Leakey Centennial Celebration, Meave still remembered piecing the skull together as "a really fantastic experience."

## RICHARD'S FINAL GIFT TO LOUIS

Even though Meave was not completely finished, Richard was anxious to show 1470 to his father, who was leaving the next week for England. On September 26 he flew to Nairobi with 1470 to show Louis what was then one of the most complete skulls found. Mary was there and the three of them enjoyed a day of closeness and conversation that Mary characterized as "almost like old times" (M. Leakey, 1984, p. 159) before Richard drove Louis to the Nairobi airport. Louis was thrilled and felt Richard's find of 1470 vindicated him and would prove to the world he had been correct for the past half century when he proclaimed East Africa the birthplace of a large-brained "true man." *Homo habilis* had tripled the known age of early humans (Hellman, p. 167), and now Louis's son would raise it to over 2.6 million years. Louis also told Richard to expect that, as in the case of Louis's finds, colleagues would raise questions. Louis was right about an onslaught of doubts from colleagues, and the 2.6-million-year-old date was wrong by over half a million years. Although Louis always relished a good fight, he was not part of the furor that followed.

## LOUIS'S DEATH

In the fall of 1972 Louis felt better than he had in several years. In May 1971 he had fallen at a conference in San Francisco and doctors drilled into his skull to remove two blood clots. Amazingly, his entire condition, including walking and finger movement, improved. He was also more rational and his family and friends believed that some of his bizarre behaviors and paranoia might have been induced by a blood clot resulting from the bee stings.

When Louis left Nairobi September 26, his blood pressure was extraordinarily high, although he was under less stress after persuading his son Colin to leave Uganda. Louis correctly anticipated the danger posed by the dictator Idi Amin and Colin was one of the last Europeans to escape from what turned into one of the bloodiest regimes in African history, with up to 400,000 deaths and disappearances.

Louis flew to London, where Vanne Goodall helped him polish the manuscript of *By the Evidence*, his autobiography that picked up where *White African* left off and covered the years from 1932–1951. Lack of energy sent him to the physician who cared for him two years before, but tests didn't indicate a heart problem; however, the doctor persuaded Louis to postpone his return to Kenya until October 5. Louis collapsed and died in the Goodall flat on October 1, 1972, at age sixty-nine.

## LOUIS'S LEGACY

Louis's contributions to paleoanthropology are difficult to quantify. That Louis made it a high-profile field of science and inspired many young people to head off to look for the "missing link" between apes and humans is inarguable; "Leakey's experience was proof that a man could make a career out of digging up fossils" (Johanson & Edey, 1981, p. 98). But his influence extended much further. Mary Smith, then a young writer for the National Geographic Society, was charmed by a conversation with Louis while they sat under Olduvai's nighttime stars. He eventually shut his eyes and told her that, with the immense resources of the National Geographic, the organization must do something to help people understand the story of evolution. "You can't have those Bible people hold back the knowledge we're learning. It must be brought to the public." Mary promised him that she would, and twenty-three years later she produced an issue of the *National Geographic* whose theme was human origins (Morell, 1995, p. 237n). Multiply this by the number of people Louis mentored, encouraged, supported, excited, and motivated to study or just to learn more about the field, and a picture emerges of a man whose immense influence reached infinitely farther than anyone can ever know.

Louis did often go over the top in describing his finds, suggesting and even advocating unsupportable theories, and touting himself. Despite these negatives, the work of Louis and his contemporary colleagues changed the focus of prehistory. Although many others contributed, Louis's showmanship and brazen self-confidence at the risk of being wrong had a great deal to do with altering the view of human prehistory. When Louis graduated from college, Asia was thought to be the birthplace of humans. By the time he died Africa was considered the continent where human history began and the date was moved to several million years earlier. A quote by Alexis de Tocqueville, born almost a century before Leakey, summarizes Louis's legacy: "History is a gallery of pictures in which there are few originals and many copies." (http://www.worldofquotes.com/topic/History/3/)

Louis Leakey was an original.

## LEAKEY LUCK

Even though they had searched unsuccessfully for twenty-five years before hitting pay dirt, the term "Leakey luck" became a catch phrase during the 1960s when Louis and Mary made important finds and *Leakey's Luck* was also the title given a 1975 biography of Louis that Mary commissioned Sonia Cole to write. Louis was undoubtedly accurate in analyzing "Leakey luck." He wrote to Le Gros Clark with a simple response: "We have been working *continuously now since February with adequate funds* and a huge labour staff" (as cited in Morell, 1995, p. 209). They would, he went on to stress, have enjoyed the same success in past years with the same resources. Mary also noted with a sense of wonder the radically increased hours of work possible at Olduvai when funding improved. More hours searching and more money for supplies and equipment translate into more fossils found.

"Leakey luck" has also been tagged on succeeding Leakeys. After all, by their second season Richard and his team had found a cache of stone tools and a variety of hominin fossils, including a complete Zinj-like skull and a second, less complete cranium. His finds earned recognition from colleagues and admiration from an interested public. Rather than luck, what Richard lacked in the way of degrees he made up for with boldness, self-confidence, creativity, hard work, and lifelong experience fossil hunting. He blended these to win his quest for success.

Richard's reasoned strategy involved reporting on the Koobi Fora finds in journals such as the American *National Geographic* that were respected but not scholarly. His colleagues then followed up with articles that listed him as coauthor in research-oriented journals like the *American Journal of Physical Anthropology* and with books like the scholarly 457-page tome *The Nariokotome* Homo Erectus *Skeleton* (Walker & Leakey, 1993). The Leakey name gave him an entrée to the scientific world, he was articulate and excelled at organizing expeditions, and the sheer volume and high quality of specimens found at Koobi Fora helped him, minus PhD after his name, become a powerful force in the field of paleoanthropology.

As Louis's life ended, Richard's professional life began. When the 300-plus bits of the three-dimensional jigsaw puzzle called 1470 were pieced together, Richard presented it at the Zoological Society of London conference honoring the 100th anniversary of esteemed British anatomist Sir Grafton Elliot Smith. Lord Zuckerman, the host of the event, congratulated Richard as "an amateur and not a specialist" and thanked him and his father for their work "not as geochemists or anything else, but just as people interested in collecting fossils on which specialists can work" (cited by Lewin, 1997, p. 164, and by R. Leakey, 1986, p. 152). After these backhanded compliments, the morning session closed. Reporters and photographers mobbed Richard and the event turned from centenary into incendiary when newspapers paid more attention to Richard Leakey and 1470 than to commemorating Zuckerman's mentor. Yes, his father's son had arrived on the scene and 1470 did for Richard what Zinj did for Louis.

Although Richard worked around his shortage of academic credentials, he had to rely on experts he trusted to date specimens. Just as Louis was blindsided by poor dating of the Kanam mandible, Richard insisted for eleven years that KNM-ER 1470 was a 2.6 million-year-old *Homo* skull, until he finally was forced to accept an age of only 1.87 million years. How did this dating problem develop? The answer to this question reveals much about the contentious nature of those in the field of paleoanthropology.

In 1969, after Kay Behrensmeyer found stone tools at Koobi Fora, the site was named after her and became known as the KBS tuff. Richard found other fossilized animal bones and tools less than a mile away. He quickly commissioned two British experts in geochronology, Frank Fitch and Jack Miller, to use the most reliable and accurate, even though expensive, tests for dating the tuff.

## DATING A TUFF IS TOUGH

A tuff is rock formed from volcanic ash. Compare a tuff to a sandwich. If lettuce and a slice of tomato are put on a piece of bread and then cheese and various types of meat are layered and finally the pyramid is topped with another slice of bread, each layer can be seen, identified, and even separated. A tuff is made of discrete layers in which ash falls on layers of rock and is covered by sediment and then the process repeats itself again and again. Reliable dates are obtained by working downward through the separate layers.

Now suppose the sandwich is made by putting peanut butter and jelly between two slices of bread. Jelly usually melts into bread and peanut butter so that the ingredients are not easily separated. A tuff can have the same problem. If a river carries ash before depositing it, the resulting tuff is contaminated. Separating the layers of the tuff to obtain a date can be as difficult as separating peanut butter and jelly from bread—and has a lot more in the way of reputation and emotional baggage resting on it.

Several methods exist for dating a fossil. In the early twentieth century, absolute dating methods did not exist and a date was established by an estimate of age combined with an educated guess. Hand axes, for example, could not be accurately dated, so they were thought to be no older than 200,000 years, since the earth was believed to be only 65,000,000 years old rather than today's several billion years.

Before Louis Leakey, few organizers of expeditions recognized the need for a geologist. Now no reputable team would take to the field without one. "Without a geologist, you really are stuck," says Louise Leakey (personal communication, October 10, 2003). At Olorgesailie, for example, what appear to be steps up a hill are cuts made by a geologist for identification of soil layers. A geologist offers valuable assistance and now uses technology to obtain more accurate dating than educated interpretation; that is, the eyeball approach.

Carbon-14 dating, developed soon after the end of World War II, is not useful for dating fossils more than 70,000 years old and in the 1970s the technology made it useful only for fossils about half that old. For materials far back in time, potassium-argon dating is used. This procedure measures the radiation from potassium found in volcanic ash to obtain a ratio of potassium to argon. In fresh ash, potassium is high. Over time argon builds up and radioactive potassium very slowly decays. This technique measures the ratio between the two and is therefore useful for dating very old fossils interspersed with ancient volcanic ash in an area like East Africa, with much past volcanic activity. Potassium-argon dating does have a greater error of measurement than carbon-14 dating, but what's plus or minus 20,000 years for material several million years old?

## DESPERATE TO GET A DATE

Miller and Fitch proposed using a refinement of potassium-argon dating on the KBS tuff—the argon-40/argon-39 technique. In 1969 it was a new and more sophisticated test that used smaller samples to yield a more reliable date. It was also much more expensive, but Richard gave the go-ahead when he was told that the date obtained would be "incontrovertible" (Lewin, 1997, p. 194). Miller and Fitch calculated the tuff to be 2.6 million years, plus or minus 0.26 million years, which certainly has a reliable ring to it. That date was later modified to 2.42 million years, still a very early date, one Richard was delighted to get because it would help fund-raising. Three years later, when 1470 was found below the area tested, the assumption was that it was even more ancient. Unfortunately, over the years Miller and Fitch stuck by their results even though they were never able to replicate the date.

What happened during the eleven years from 1969 to 1980? Almost all the major players in paleoanthropology got involved and each of them had their own personal and strongly held view. One in particular surfaced who still plays an important role in the Leakey story. Enter Tim D. White, now professor in human evolutionary studies at the University of California, Berkeley, and considered "one of the profession's ablest morphologists" (Lewin, 1997, p. 172).

Tim White worked for several seasons with both Richard and Mary when he was a doctoral student. By the mid-1970s concerns about 1470's age were widespread when Richard asked White and John Harris, then head of the paleontology department at the National Museum and Richard's brother-in-law, to study fossil pigs from the same section of soil. At first, White and Harris felt the dating was accurate. However, the longer they investigated, the more they questioned the 2.4-million-year date.

## "WHOSE CAMP ARE YOU IN?"

Finding fossils requires a well-organized camp. The word *camp* can also refer to those on a particular side in a dispute. As Roger Lewin wrote in *Bones of Contention*, anthropology is "a science that is often short on data and long on opinion" (p 64). So it was that by February 1975 those who attended a meeting of the Geological Society of London clearly identified themselves as belonging to either the Leakey camp or the critics camp. The controversy dragged on. Conferences and meetings were fueled by debate about the date, no matter what else was on the agenda. Articles were published in scientific journals. By this time the controversy was more a matter of politics than a search for scientific truth and "passions were clearly running deep" (Lewin, 1997, p. 227).

The KBS controversy spilled into the relationship with his mother and Richard and Mary were constantly arguing. Mary didn't hesitate to tell colleagues she thought Richard's team was making huge mistakes and he was unhappy when someone passed on the condescending remarks his mother made. Richard quietly cut her from his life, as he did when situations flared with other family members. He later said she was "dismissive of many of the things I did" (as cited by Morell, 1995, p. 425), with a patronizing attitude toward his work at Koobi Fora, and he coined the word *matronizing* to describe her unwelcome meddling.

In 1976 Harris and White shared the final draft of the fossil pig report with Richard before sending it off for possible publication. It included a date of less than two million years and Richard was upset, not, he said, because it didn't support his date but because they referred to several fossils about which nothing had yet been published. Richard ran a very tight camp and forbid others writing about a fossil before the team member who found the fossil had published. White was furious and slammed the door as he left Richard's office. White and Harris rewrote the article and Richard assumed the issue was closed but White vented to others about what he saw as censorship.

At last, in 1980, influenced by growing evidence that included the rewritten fossil pig report from White and Harris, arguments presented by his friend Glynn Isaac, and dates using both potassium-argon and argon-40/argon-39 tests obtained by two other labs, Richard realized that he should and could no longer support Fitch and Miller. He had recognized for several years that the age could be wrong but admitted to "not knowing anything about geophysics and dating" (R. Leakey, 1986, p. 170) and he wanted to support people he trusted. To the credit of Richard and those in his camp, the geology of Koobi Fora looks, as Roger Lewin described it, "in some places like a moonscape and in others like a bulldozed building site" (1997, p. 235) and dating such soil is horrific. Richard noted that he regretted losing several friends because of the controversy, but improvements in dating resulted and the KBS tuff is now "one of the most securely dated deposits in Africa" (R. Leakey, 1986, p. 170).

## ACTION-ORIENTED

Except for the KBS controversy and Richard's declining health— admittedly two huge exceptions—Richard's life was busy and productive during the mid-1970s. In June 1974 Richard and Meave welcomed the birth of their second child, Samira. Both girls spent time under the care of a nanny although Meave balanced career and family by working at the museum rather than in the field. She considered her relationship with her husband and caring for her daughters important. "If you're going to have a family, if you're going to have children, don't try to make your career No. 1 in your life. . . . They grow up fast, and they're very special" (cited by Strahle, 2001). Both children were involved in fossil expeditions when they were old enough to be useful. Louise enjoyed learning to drive at age twelve by navigating the bumpy Koobi Fora roads in a Land Rover, but dinners were often boring to the girls, with long names like *Australopithecus* thrown around a great deal.

Also in 1974 Richard established the Foundation for Research into the Origin of Man (FROM); made a documentary for British and American television; raised money to honor Louis with a prehistory wing at the National Museums of Kenya, where a statue of Louis flanks the entrance; worked with several wildlife and environmental organizations; and wrote two books with Roger Lewin. *Origins*, which explains the history of earth in a conversational style, sold half a million copies and was published in ten languages.

Involvement with other activities did not stop the flow of fossils found at Koobi Fora. Richard's team had rightfully earned the name "Hominid Gang" for their ability to spot fossils. Kamoya Kimeu was the team leader and notified Richard about finds. Richard had a camp rule that no hominin was to be dug until he inspected it, collected data, and considered the best technique to use before he excavated it.

On August 1, 1975, Bernard Ngeneo, whose keen eyes had first seen a bit of 1470 amid deposits of small pieces of bone, spotted a hominin brow ridge. Recognizing its potential, Richard carefully took his time with excavation. At night they placed a metal pan upside down and some branches from a thorn bush over the gradually exposed skull to protect it from animals wandering through. The digging was filmed and Meave recorded the progress in her diary until Richard finally held a 1.75 million year old skull—KNM-ER 3733—securely in his hands on August 9. At the time regarded as a *Homo erectus*, most paleoanthropologists now consider it the distinct but similar species that moved out of Africa to populate the world, *Homo ergaster*. The skull was very complete and some teeth were found in sieving the nearby soil. "What a find it was!" (R. Leakey, 1986, p. 172).

Unfortunately, the skull was like an eggshell over calcified rock that made it ponderously heavy and difficult to examine. Alan Walker, a close friend and paleontologist based in the United States, had a plan. He would cover the skull with tissue paper and glue, literally turning it into a weighted piñata. Rather than people with sticks swinging to break it open, he would use a chisel at its base to crack it. Voila! The skull would pop off the rock in hopefully very few pieces. Richard elected not to be anywhere near when Alan, with help from Meave and Tim White, carried out his plan. After frustration from fruitlessly bashing, they obtained a heavier chisel and the skull did indeed break away from the rock in three pieces, ready to be cleaned and perfectly glued together.

KNM-ER 3733 represents an adult *Homo ergaster*, an early species of genus *Homo*. KNM-ER 3733 is probably female; the identification of sex comes from a comparison of her face with another Koobi Fora cranium, KNM-ER 3883, and with KNM-WT 15000, a male later found on the opposite (western) side of Lake Turkana. The features of KNM-ER 3733 are markedly less robust, but fully closed cranial sutures, wear on the teeth, and third molars indicate that she was an adult at the time of death and not a child. From what they deduce about a specimen, scientists can then use the similarities and differences among fossil bones to understand relationships and make assumptions. For example, since *Homo ergaster* was able to trek from Africa and become the ancestor of later *Homo* populations, this species would seem able to adapt to different ecosystems.

## STEPPING OUT

Mary's personal evolution during Louis's last years paved the way for professional independence. An introvert, she was more at ease in the field than dealing with the public and was dedicated to scientific precision. In a *Newsweek* (October 29, 1984) review of Mary's autobiography, *Disclosing the Past*, Sharon Begley wrote, "Although Mary's growing independence from Louis contributed to the breakdown of their marriage, it also helped her become a leading scientist in her own right after his death in 1972" (p. 119). The stage was set for footprints in the sand, or, to be more exact, in the volcanic ash.

Way back in 1935 an unknown African visitor, half Maasai and half Kikuyu, encouraged Louis and Mary to visit a location called Laetoli where he had found bones that were like stone. When he returned for a second visit with samples, they decided to let him guide them there. Although today a brief trip south, reaching Laetoli from Olduvai in 1935 was a circuitous two-day drive, but their guide easily located the remote area. (In her autobiography Mary compared Richard's twenty-four-hour rapid reconnaissance of Lake Turkana by plane with this first inspection trip she and Louis made to Laetoli [M. Leakey, 1984, p. 147]). Amidst deposits of sediment containing volcanic ash, they found a number of fossils, including one eventually tossed in a box of miscellaneous fossils at the British Museum. Many years later it was identified as hominin.

Despite the possibilities Laetoli offered, Mary did not begin a major expedition there for forty years. Laetoli soil was established by potassium-argon dating as between 3.59 and 3.77 million years old, a date obtained by Garniss Curtis, a critic of the KBS tuff dating. It was therefore much older than Olduvai, even though only thirty miles separated the two areas. Thus it was that finally, in 1975, at age sixty-two, Mary set up camp at Laetoli. Although they relocated the camp in 1976, the problems that had kept Mary away for so many years still made fossil finding there extremely difficult. Laetoli was so remote that bringing in supplies, including fresh water from ten miles away, was a challenge. Blistering hot days contrasted with cold nights to make the weather less comfortable than at Olduvai. Buffalo, which are aggressive and charge humans, were a constant threat and poisonous puff adders were plentiful. Ticks peppered clothes and had to be brushed off several times a day. In addition to these difficulties, in 1975 Mary had health problems. Despite the problems, in the first season, Mary's team found a variety of hominin teeth and jaws. Then Laetoli produced what Mary called "perhaps the most remarkable finds I have made in my whole career" (as cited by Lewin, 1997, p. 278).

More than 3.6 million years ago, Sadiman, the volcano that overlooks Tanzania's Serengeti plains, spewed ash containing a high degree of carbonatite and, by chance, rain fell. A variety of animals, including an ancestor of modern humans, crossed this muddy muck and left their footprints, as someone today might leave their footprints in a concrete sidewalk that's still wet. An additional six layers of ash and soil covered and preserved the prints until erosion made them again visible.

On September 15, 1976, biologist David Western, having a little fun, hurled some dried elephant dung, aiming at Dorothy (Dodi) Dechant, geologist Kay Behrensmeyer's assistant. Elephant dung is shaped more like a softball-sized muffin than a cow pie, but very light, with thorns in it. This toss actually was not unexpected. Kay Behrensmeyer notes, "David Western is known to his friends for enjoying a good dung fight, especially when a long walk in the hot sun needs 'lightening up.' He has even been known to toss them out of his airplane, targeting hapless scientists on the ground!" (personal communication, November 1, 2004).

Soon muffin mayhem broke out, with ducking and scrambling as Kay, Philip Leakey, and paleoanthropologist Andrew Hill also got into the action. Behrensmeyer remembers, "Andrew and I went down into a shallow gully to protect ourselves from the barrage." To avoid being hit, they bent over and as they did their eyes chanced on animal and bird footprints preserved in the hardened ash under their feet. Both suddenly realized "the footprints of elephants and small antelope were exposed on a flat bedding plane on the floor of the gully, washed clean by the last rainy season. We both saw them but Andrew was first to remark. I am happy to give him credit for being first to articulate his find" (A. K. Behrensmeyer, personal communication, November 1, 2004). And what a find it was! The footprints provided a remarkable window into the past of almost four million years ago. Thousands of tracks were eventually uncovered of animals that exist today, like giraffes and monkeys, and those of some long extinct, like a three-toed horse.

As the discovery of the trail of prints slowly surfaced over several years, their importance also evolved. At the end of the 1976 season, Peter Jones, an archaeologist who was Mary's assistant for many years, and Philip Leakey found prints that seemed as if they might be human, footprints "so extraordinary" Mary could "hardly take it in or comprehend its implications" (M. Leakey, 1984, p. 177), until finally in 1978 the team felt they had unequivocal evidence that early humans walked on two legs across the Laetoli landscape. A 1974 discovery of an australopithecine named Lucy confirmed what some paleoanthropologists had theorized for a number of years—that early humans walked erect before big brains evolved. The Laetoli footprints provided powerful additional evidence.

The trail fascinated Mary and her team as they tried to understand how early humans made the tracks and attempt to replicate the path made 3.6 million years ago. Was one set of prints made by someone with a disability who was helped by a smaller person? They tried different styles of walking and discussed and joked about Pliocene behavior. Were two of them lovers who walked incredibly close to each other? Alan Root, a wildlife movie photographer, came up with a likely, although not universally accepted, answer: There were not two humans but three, with the third, small one, walking in the footsteps made by the largest, much as a family walking in wet sand across a beach today.

Information about the trail of footprints had been published in an article in *Nature* two years earlier but Mary, ever the cautious scientist, had witnessed the hubbub that occurred many times when Louis rashly named a new species. No, Mary would wait until the majority of the human footprints were uncovered to name the ancestor that traveled across Laetoli's soil and whose fossil evidence, which Mary felt was early *Homo*, they had found during her first season at Laetoli. Mary also planned to meet with Donald Johanson, who had found some exceptional fossils at Hadar in the Afar valley of Ethiopia, and Tim White, who was the first to notice a similarity between the Afar and Laetoli finds. Mary had been impressed by White's report on Laetoli's fossils that he did at her request so when he asked Mary to entrust him with excavation of the footprints, she did, although she considered her Kenyan workers very capable of directing the job. Mary felt White wanted to make decisions that should be hers, but they had talked through their differences and seemed to be on friendly terms at the end of the 1977 season.

Mary at last decided to announce and describe the footprints at a press conference in January 1978 when she was in Washington, DC. She would then discuss them in detail at a presentation in May 1978 at the Nobel Symposium in Sweden. There King Gustav would also present her with the Linnaeus Medal, the first woman to receive this honor, for contributions to the field of science. It was a decision that caused later anger and aggravation.

# CHAPTER 8

# CONTENTIOUS, CONTENTION, AND A CRITICAL CONDITION

*(1978–1988)*

> Science is driven by questions, not answers. The scientific arguments surrounding human evolution are intellectually stimulating. . . .
>
> —Jeremy DeSilva, "Interpreting Evidence," p. 262

## THE DONALD OF PALEOANTHROPOLOGY

In the early 1970s, Mary, Richard, and Meave Leakey and Donald Johanson shared information via mail and visited each other's camps to examine and discuss fossils. Johanson was on the board of directors of Richard's Foundation for Research into the Origin of Man (FROM), sailed with him, and was a guest of the Leakeys numerous times, including Christmas 1975. Although there were disagreements about interpretation of fossils, they considered themselves friends.

The scene shifts to the May 1978 presentation at the Nobel Symposium at the Royal Swedish Academy of Sciences. Donald Johanson was a rising star in paleoanthropology. He was at the symposium to describe unearthing a partially complete skeleton and approximately 200 fragments of thirteen or so individuals he found near Hadar, in the Afar region of Ethiopia, that he called his "First Family." He remembered the event this way:

I mean, here was assembled fifteen of the world's specialists in human evolutionary studies. Richard Leakey was there, Mary Leakey was there, a whole host of people, from prestigious universities, who were published widely, and here I was. 1978, I was at that time a young scholar, thirty-five-years-old, making this announcement. And furthermore, I presented a new view of how the family tree looks. (Academy of Achievement, 2004)

Johanson had been born into a middle-class family of Swedish immigrants in Chicago, Illinois. In 1973, when he was thirty-years-old and still working on his doctorate, Johanson went to Ethiopia as one of the leaders of the Joint International Afar Research Expedition. Four years later he had amassed an amazing collection of hominins that included almost half[1] of a skeleton he called Lucy. The selection of its popular name was both brilliant and seemingly serendipitous. Johanson and his team heard the Beatles' song "Lucy in the Sky with Diamonds" playing on the radio the night of the find. Easy to remember and easy to fit into a headline or sound bite, Lucy was a diamond-studded choice of name and made Johanson a glowing star in the paleo-sky. Americans were happy to embrace one of their own as the new superstar of paleoanthropology. Reactions at the Nobel Symposium, however, weren't what Johanson expected.

I thought that this was going to generate enormous discussion. I finished my paper, and there was a question and answer period, and nobody asked a question. They broke for tea, people left the room, and only one scientist came up to me afterwards, and said "It's unbelievable." They were so taken aback by this that they didn't even want to discuss it. During the week's discussion, whenever people would start debating a family tree, I would say, "What about my family tree? What about what I've suggested?" Some people deliberately tried to ignore it, and not consider it, because it really upset their whole view of human origins. (Academy of Achievement, 2004)

Why did Johanson's words take people aback and upset them? Partly because the implications of what he said were a great deal to unravel and comprehend but also because some in the room felt angry.

Johanson was guilty of a breach of etiquette far greater than using the wrong fork at dinner. In fact, he was guilty of three breaches of etiquette: grouping Lucy and Laetoli into the same taxonomic group, choosing the Laetoli fossils as the holotype (or type specimen with which similar fossils are later compared), and scooping Mary. These three are scholarly issues, but Dr. Cathy Willermet, University of Texas at El Paso, notes that paleoanthropologists become attached to their finds and "the issues are intellectual but they are also emotional" (personal communication, November 29, 2004).

In explaining his extraordinary find of Lucy and the "First Family," Johanson also talked about Mary Leakey's Laetoli footprints. In fact, he ascribed them to the same species, a new one he and Tim White named *Australopithecus afarensis*, southern ape-man from the Afar. Mary knew Johanson was going to present the name, which she did not like. When Donald Johanson and Tim White lumped Lucy and the First Family along with the Laetoli specimens, it represented a general reorganization of hominin evolution. Mary and Louis never saw australopithecines as human ancestors and she did not want her Laetoli finds to be designated as australopiths. "Call it what you like, call it *Hylobates*, call it *Symphalangus*, call it anything, but don't call it *Australopithecus*," Mary said (as cited by Morell, 1995, p. 490). She also pointed out that by linking Lucy with the well-dated Laetoli site, Lucy became the oldest yet known fossil hominin—and to the finder of the oldest fossil goes fame and funding. Mary Leakey's most extraordinary find will always bear the notation that the footprints were made by *Australopithecus afarensis*, even though the biped who made them lived far from the Afar, about 1,000 miles to the south, and was separated from Lucy by half a million years.

Second, Johanson and White didn't choose Lucy or one of the First Family as the holotype. They chose, as is common, the first fossil found to represent the new species, in this case a broken jawbone minus some teeth from Mary's Laetoli collection. Johanson later wrote that they thought including Laetoli would please Mary (Johanson & Edey, 1981, p. 289). It didn't. Roger Lewin explained, "Mary Leakey saw 'her' fossils being appropriated by Johanson and merged into anonymity with the Hadar fossils. And she did not like it" (Lewin, 1997, p. 171).

And last, Mary didn't realize Johanson would discuss her Laetoli footprints so thoroughly (Lewin, 1997, p. 290). Etiquette forbids a scientist from formally discussing another scientist's fossils until the discoverer of the fossil officially presents the information, which then becomes a matter of public record and open to all for comment. In Johanson's view, Mary Leakey had already published information about the footprints and discussing her find within his presentation was proper because his comments provided crucial context for Lucy. Mary, however, was the speaker scheduled immediately after Johanson and she felt he made a fool of her when he both scooped her and relegated her find to the background.

Mary was not the only speaker who was upset at being scooped. Phillip Tobias, in trying to mediate, planned to untangle the Laetoli and Hadar confusion by naming each as a subspecies of *Australopithecus africanus*, but Johanson scooped him along with Mary.

Mary may have fumed but she returned to Laetoli and by 1979 had completed excavation of the footprints. They show amazing similarity to today's human foot, with a defined arch and a big toe in line with the other toes, as opposed to the long free great toe of a chimp. Based on analysis of Lucy's bone structure, they also indicate that *A. afarensis* probably did not walk exactly the way humans today do, on two legs all of the time. There are competing theories as to whether *A. afarensis* showed bipedal behavior but lived or slept primarily in trees or whether its posture was upright most of the time (Tattersall, 2002, pp. 87–89).

Despite contributions and attempts to build a museum at Laetoli, Mary decided the location was too remote for visitors and a museum too difficult to protect year-round. Casts were made of the entire trail and the footprints were reluctantly again covered.[2]


## HIS FIRST LIFE ENDS

After 1470 threw Richard onto the world stage, he remained there by writing best-selling books and making lecture tours that, like his father's, led to funding. Despite denial of his health problem, by late 1979 Richard looked puffy and bloated and he could no longer deny his situation was serious. Dialysis for the rest of his life where electricity and sanitary conditions can be iffy, as at Koobi Fora, meant a machine would control his comings and goings. A transplant was the only solution. He was so ill by the time he arrived in London that the specialist would not allow him to return to Kenya to see his home and children but insisted he remain there to await surgery.

Surprisingly, considering the acrimony that existed in the family, Richard's two brothers, Jonathan and Philip, offered him a kidney, as did his half-brother Colin and several friends. Philip proved the best match but asking him to live the rest of his life with only one kidney was unsettling to Richard. He and Philip had been estranged for over a decade and had never been close since childhood. Both left home when they dropped out of school but the similarities ended there. While Richard seriously pursued success, Philip was "wild and carefree" (Morell, 1995, p. 510). He lived with various tribes, tried his hand at all sorts of disparate jobs, made money and lost money, and lived as he pleased. All communication between them had ended in 1969 when Philip refused to move from land that Richard wanted to add to a national park (Morell, 1995, p. 510).

Richard finally asked and Philip agreed to part with a kidney, but he was running for a seat in the Kenyan Parliament and wanted to wait until after the election, a date that unfortunately was postponed several times.

Since Philip seemed to be settling down and finally having a focus in his life, Richard and Meave readily agreed to wait, although Richard's health was deteriorating rapidly. As a Kenyan, he didn't immediately get access to a dialysis machine and, as he did when he was in Kenya, he assured the doctors he was fine and could wait. Only Meave realized how severe his condition was. Soon walking up a flight of steps was almost impossible and he spent most of his day lying on a sofa, nauseous, cold, weak, and tired. Dialysis at last became available and was a tremendous boost. Louise and Samira came to London to make life seem more normal while the family waited for Philip to arrive.

At last mid-October elections made Philip the first white Kenyan to be elected to Parliament since the country's independence from Great Britain. Three weeks after the election, Philip and his wife, Valerie, quietly left for London. Like Richard and Mary, both of whom also quietly traveled to London, Philip did not want publicity about his brother's health or his contribution. At the first meal Philip and Valerie ever had with Richard and Meave, a rather uncomfortable luncheon meeting, Philip turned down a chance to back out of the operation. However, he did insist on protection from security police because his life had been threatened after the election and he had information that an assassination attempt would be made in the hospital. With only four days till the scheduled surgery on November 29, 1979, Richard quickly and unhappily called friends in London and bodyguards were arranged so the operation could be performed as planned.

The surgery was much more involved for the one who donated, so on their gurney trip to the operating room, Richard offered Philip one final chance to renege. The nurses laughed when Philip's reply was "short, blunt and unprintable" (R. Leakey, 1986, p. 201). That night Richard awoke to what he hoped to see: a bottle "containing a rather bloody liquid which I immediately recognized as my urine" (R. Leakey, 1986, p. 201). Dark humor replaced anxiety. One favorite joke was that Richard could no longer say he hated Philip's guts because he had part of them.

Richard returned to the hospital less than a month later. Drugs stopped his body's rejection of Philip's kidney but they also suppressed his immune system and back to the hospital he went yet again, this time with life-threatening pneumonia combined with septicemia, a type of blood-poisoning. During the twenty-four hours he hung between life and death he had an out-of-body experience in which he felt at peace and wondered why he should return to his body. But then he saw Meave holding his hand and heard her encouraging him to breath and fight. He later credited the efforts of the hospital staff and especially Meave with saving his life (R. Leakey, 1986, p. 202). Eight months after his disheartening flight to London, when he didn't know if he'd ever see home again, the family returned to Kenya.

Sharing a kidney didn't solve the personal differences that existed between Philip and Richard and soon the brothers were again no longer speaking. In the 1980s, what Richard considered purely political reasons motivated Philip to speak out against a mutual friend who had once helped him. He urged Richard to also denounce him, but Richard refused to abandon a friend. The two brothers are also at opposing positions in Kenyan politics.

## RICHARD REBOUNDS

When Richard returned to Kenya in February 1980, the political situation at the memorial institute he had built for his father was messy, but Richard merged this institute with the National Museums, despite criticism from home and abroad, and began to juggle his job as director of the museum with a BBC television series and new expeditions to Lake Turkana. For good measure he started a foundation to help kidney patients in Kenya.

## JOHANSON: REPRISE

Within months after his return from London, Richard received an advance copy of *Lucy: The Beginnings of Humankind*, Donald Johanson's 1981 book written with Maitland Edey. Richard and Mary were infuriated about what they felt were obvious inaccuracies in reporting stories about Louis and their family. The book also paid them a number of compliments, but they no longer viewed Johanson as a friend. Up till now Richard felt disagreements with Johanson were more academic than personal. He had continued to exchange friendly letters with Johanson and even helped him obtain a National Science Foundation grant; now he suspected Johanson felt fame outweighed friendship.

Johanson had not succeeded in unseating Richard from his perch atop the world of paleoanthropology although newspaper and magazine articles, especially those from America, helped by building up a rivalry between the two. Johanson was usually cast as David and the Leakeys, past and present, as Family Goliath. The final battle came in the spring of 1981 while Richard was in New York for a FROM board meeting. His publisher interrupted the meeting to urge him to join Johanson in what was supposed to be a televised discussion about creationism and evolution. It wasn't.

Richard had been apprehensive about appearing on national television without advance preparation, but he is a polished presenter and his publisher urged him to join Johanson in what Richard was assured would be a dialogue. When Richard arrived at the television studio, he soon realized that the stage was set for a debate about Lucy and the differing opinions each of them had about human origins. Johanson was well prepared and equipped with props. Richard had no casts of skulls, since he rushed directly from his meeting, and he was immediately ill at ease and felt he had been trapped. With the cameras rolling, venerable CBS newsman Walter Cronkite set the stage by saying that Johanson had found Lucy and, in doing so, passed Richard in the race for the "missing link"—fossil evidence that connected people today with apes past. With Johanson showing and explaining casts of his fossils, Richard felt at a disadvantage and was far from his usual articulate self. As the program ended, Richard drew an X over Johanson's version of the human family tree and a big question mark to the side to show that the fossil record still needed to be filled in. It was not the finest moment for either of them; it was the last time they spoke to each other.

John Van Couvering, at the end of a book review about fossil hunting, wrote, "I recall the wry diagnosis of 'the hominid game' offered by a paleontologist who was in Kenya studying fossil fish at the time. 'It's always a bad combination,' she concluded, 'when you get hominid fever on top of testosterone poisoning'" (Van Couvering, 2001). Both Donald Johanson and Richard Leakey had caught the fever at an early age.

Although the years have mellowed both sides somewhat, collaboration between Johanson and the Leakeys stopped and antagonisms still exist.

## TURKANA BOY

On a late summer evening in 1984, Richard received a phone call. It was the leader of the Hominid Gang, Richard's trusted friend, Kamoya Kimeu. Kimeu had found a small, 1-1/2 by 2-inch piece of hominin skull along a slope leading down to a narrow gully. It seemed to be a very unpromising site and one that had been searched in the past, and Richard wrote in his journal, "Seldom have I seen anything less hopeful" (R. Leakey & Lewin, 1993, p. 34). The next morning they first visited another site before moving to the site Kimeu had found. In the afternoon, bored by sieving at Kimeu's site, Richard and several others left and returned in the late afternoon to hear shouting for them to look at the bones they had found. "Lots of skull!" (R. Leakey & Lewin, 1993, p. 40). Kimeu later said he had a sense that he should look near the gully and Richard, having had similar premonitions, understood completely. Soon the camp was crowded with visitors and anticipation was in the air. Alan Walker had washed the bones in the camp shower and begun to put together the 1.6 million year old jigsaw puzzle.

The site was close to their camp near the Nariokotome River, an easy walk no further than the length of three football fields. Although by dusk on August 29 they found a number of bones, they still had doubts they would net much more and after a few hot, discouraging hours the next morning they were ready to quit. To add to their difficulties, Richard was digging in the roots of a wait-a-bit thorn bush and they were aggravated by thorns hooking their clothes. Suddenly Richard and Peter Nzube spotted, protected by the tree's roots, a "perfect half upper jaw" (R. Leakey & Lewin, 1993, p. 44). Ten-year-old Samira arrived with more news from Alan Walker and Meave. Her note said that the bones were those of a child.

How did Alan and Meave know? They counted the skull's teeth and realized the third molar on the right side was missing. They also compared the skull's teeth with those of twelve-year-old Louise and deduced that the individual had died at about her age. During the next three weeks, a period Richard called "paleontological bliss" (R. Leakey, 1993, p. 45), they excavated vertebra, clavicle, and so many other bones that they became almost blasé about bringing into sunlight the first bone of its type modern eyes had ever seen. Until their find, only fragments of *Homo ergaster* had been discovered. (Recall that *Homo ergaster* is the term used to designate *Homo erectus* before it moved out of Africa).

Turkana Boy, KNM-WT 15000, is a nearly complete skeleton and a prime example of what bones reveal about early ancestors and their way of life and how much can be deduced from minimal facts. The length of Turkana Boy's thighbone indicates he was five feet, three inches tall at age twelve, and would have stood over six feet tall by the time he was an adult, much larger than other species that had evolved by about 1.54 million years ago. His narrow pelvis tells paleontologists that he was slim and his large body surface suggests he perspired to cool himself in the tropical heat of Africa and probably walked speedily on his two long legs. To an untrained eye, his rib cage appears identical to humans today but investigation from the neck up reveals how very different his lifestyle was. His lack of forehead shows small frontal lobes, the part of the brain required for complex thinking. With a brain the size of today's one-year old, he was unable to reason, but Alan Walker calls him "devastatingly clever for his time" (Gore, 1997, p. 91), since his brain was twice the size of a chimp's. Of course, no record exists of his consciousness—his thoughts, feelings, and sensitivity toward others.

Turkana Boy lived with others of his species but probably communicated only through very basic language, more than signals and grunts but not language as spoken today, although speech is a complex issue and its development open to dispute. The part of his brain that controls vision is large, suggesting that his sight was excellent and that he hunted animals as a food source, especially since sophisticated hand axes used to butcher have been found and dated to this time period. Turkana Boy's diet must have included meat, since the teeth and jaws are not the huge size *Australopithecus boisei*, or Nutcracker Man, needed to crush nuts and pulverize vegetation. Alan Walker, then at Johns Hopkins University, showed clinicians at Johns Hopkins Hospital a slide of a section of leg bone. The clinicians identified death from hypervitaminosis A, probably from eating carnivore livers that contain huge amounts of vitamin A. They were astonished that the bone used for their diagnosis was from a child who died almost two million years ago rather than a recent death (Leakey & Lewin, 1993, p. 61).

## THE BLACK SKULL

August 1985: While his wife, anthropologist Pat Shipman, coated a hippo skull with Bedacryl, used to harden fragile fossils and make excavation easier, Alan Walker decided to look for fossils. He returned and casually asked if she wanted to see a hominin. Of course she did. He had located a cairn of stones Kamoya Kimeu built to mark a piece of skull he found several days before. Soon the crew was digging the many fragments of what turned out to be a skull dating to at least 2.5 million years. It was dubbed the Black Skull (KNM-WT 17000) for the bronzed-black color that resulted from manganese salts seeping into the bone from the surrounding soil.

Finding a hominin is difficult, but interpreting what has been found often raises very thorny issues. Deciding what they had found was more complicated. Kamoya repeatedly commented about the size of the tooth roots when he first saw the skull which, when glued together, showed a mixture of features. Richard and Alan realized that disagreement would follow whatever name they chose. They eliminated *A. boisei* because of several different and distinctive features and finally settled on *A. aethiopicus*, a name that had been given a fragment of a jaw way back in 1967, oddly enough by the French team of the Omo expedition that Richard left to go his own way at Koobi Fora. "The issue was debated, but today there is general consensus that there was a distinct robust species, *Paranthropus aethiopicus*, living in eastern Africa 2.5 million years ago" (Smithsonian Institution Human Origins Program, *Paranthropus aethiopicus*). Although, to no one's surprise, there is much debate how *P. aethiopicus* fits into the picture of humankind's beginnings, this species seems like a probable ancestor to *P. robustus*, originally found in southern Africa in 1938 with later finds in 1948, and *P. boisei* (Zinj) (Smithsonian Institution Human Origins Program, *Paranthropus aethiopicus*, and Tattersall, 2002, p. 127). Other paleoanthropologists, including Tim White, consider KNM-WT 17000 to belong to the *Australopithecus* family (DeSilva, 2004, p. 259). Designations represent differing interpretations of an ever-expanding fossil record.

## MARY, LEAVING THE FIELD TO OTHERS

Mary continued to live and work at Olduvai but she traveled to speak at fund-raising events and to receive a number of awards and honors. When she received an honorary doctorate in 1981 from Oxford she described the long ceremony, with much Latin spoken, "as if they understand it" (M. Leakey, 1984, p. 198). After she took the stage and survived "a further brief outbreak of Latin" (p. 199), she sat on a reserved chair and tried to find friends' faces among the audience. The evening was not pleasant. Mary didn't feel well and found the speeches after the formal dinner condescending to women. Then only male guests were offered cigars, which Mary loved, so she appropriated one, "doubtless to my hosts' disapproval" (M. Leakey, 1984, p. 200). She was relieved when the evening was over.

No matter where she was, Mary was always outspoken and a person of extreme likes and dislikes. Peter Jones was driving when Rick Potts, then a young paleoanthropologist, made his first trip to Olduvai. Jones stopped at the gate and told Potts to open it and expect to be greeted by Mary's Dalmatians. Mary would be watching from the house, Jones said, and how the dogs responded to Potts mattered to her. Potts, who was a little fearful of dogs, hesitated slightly as he opened the gate and saw the large barking animals running toward him. But he reluctantly let them jump on him and lick his face. It obviously paid off. Mary loosened up more with Potts than with many people, shared stories with him over the years, visited his digs at Olorgesailie, and always considered him her friend. Potts benefited by losing his fear of dogs (R. Potts, personal communication, October 26, 2004).

In 1983 Mary left Tanzania. Travelers had once casually moved between Kenya and Tanzania; now the border was closed and, although she and her team could travel without difficulty, her family was unable to easily visit her, so life there was lonely. Even the lions and rhinoceroses, almost made extinct by poachers, were no longer there for company. Tanzania was her adopted home and had been her refuge when marriage to Louis no longer was, but it was not an easy place to live. She considered George Dove her good friend and neighbor, but he lived twenty-five miles away.

A blood clot behind her left eye that left her blinded hastened her move to Nairobi. Meave welcomed her at their home, and Richard himself cooked gourmet meals for her until she was able to move into her own home at Langata. The house was modified so two rental properties provided a steady source of income and gave her adequate living quarters.

Moving from Olduvai must have been very difficult for Mary, but Rick Potts feels her entire personality changed when she returned to the world outside (R. Potts, personal communication, October 26, 2004). Now there was time to visit friends, get to know her nine granddaughters and one grandson, write her memoirs, complete an analysis of excavations at Olduvai, and return to her old love of prehistoric art.

Mary's father had influenced her broad intellectual interests and led her toward a special attraction for cave paintings in France, which, as an artist and a lover of all things related to ancient cultures, he particularly relished. In the typical way that the Leakeys combined holidays with prehistory at a different location, Mary visited caves in France and Africa over the years and laboriously copied the artwork. Since drawings were layered on top of other drawings on the cave walls, the only way to sort out the various figures was labor-intensive tracing of each figure, ignoring the parts of the drawing that were not relevant to that particular image. Her tracings illustrating early people and animals were stored in a trunk and it was not until 1983 when, with encouragement from daughter-in-law Meave, *Africa's Vanishing Art* was published. Waiting paid the dividend of using more modern techniques to reproduce color-correct illustrations of early animals and people.

Mary also traveled some, including visits to Olorgesailie, where Potts was conducting research. When Potts and his crew found a very strange rectangular area cut into the soil near huge fossil elephant bones they were excavating, he asked her about it. It turned out to have been one of Louis's old digs. Louis had missed finding the elephant by only a few feet for which, Mary said, showing her wicked sense of humor, they should all be grateful (R. Potts, personal communication, October 26, 2004).

Again, Donald Johanson proved to be a thorn in Mary's side. Comments in his 1981 book, *Lucy: The Beginnings of Humankind*, infuriated her and she shed tears and anguished over what she considered awful lies about her and her work, even though she had opened her house to him and treated him well professionally and personally (Morell, 1995, p. 537).

In much the way she felt he had appropriated her fossils at the 1978 Nobel Symposium, she perceived that he did the same at Olduvai. When Ethiopia closed its borders to fieldwork by foreigners, Johanson, along with Tim White, moved to the Olduvai camp, which had been turned over to the Tanzanian government, and worked there briefly in 1985 and then for a full 1986 season. Johanson wrote Mary only a very brief note in 1986 in which he asked for the number of the last hominin fossil found at Olduvai. He had found a fragmented skeleton and wanted to begin his numbering in sequence. Mary replied in kind with a note congratulating him on finding what became OH 62. Again, Mary and many of her fellow scientists had problems more with Johanson's lack of manners than with his actions and felt he showed insensitivity to Mary's long relationship with the camp. His account of *Lucy's Child: The Discovery of a Human Ancestor* (Johanson & Shreeve, 1989) was the result of Johanson's brief stay there. Mary, however, had a lifetime of memories about Olduvai.

## NOTES

1. Forty-seven of Lucy's estimated 207 bones were retrieved (http://www.mos.org/evolution/fossils/fossilview.php?fid=22). Donald Johanson refers to the "completeness" of Lucy (Johanson & Edey, 1981, p. 24), but Bill Bryson (2003, p. 444) points out that Johanson did not count hand and feet bones. Since Lucy's claim to fame is her ability to walk on two feet and use her hands, not including these bones is a major omission.

2. A photograph of one of the footprints can be seen at (http://www.mnh.si.edu/anthro/humanorigins/ha/laetoli.htm. A three-minute video explaining the evidence the prints provide that humans walked on two legs almost four million years ago and comparing the Laetoli prints with human footprints today and chimp prints is at http://www.pbs.org/wgbh/evolution/library/07/1/l_071_03. html.

# CHAPTER 9

# BURNING HIS TUSKS
# BEHIND HIM

## *(1989–2003)*

> A huge pyramid of more than two thousand elephant tusks rose
> some twenty feet above us in a sort of macabre sculpture. . . . I
> then handed him a lit taper, and we walked together to the edge
> of the ivory pile. Without hesitation, President Moi plunged it
> into the kerosene-soaked straw. In a flash, flames leaped sky-
> ward, engulfing the ivory in a hot, red-orange blaze.
>
> —Richard Leakey, *Wildlife Wars*, p. 92

## A GRAND GAMBLE

Richard was relieved that the tusks, and not Kenyan President
Daniel arap Moi, lit the Kenyan sky. It was a well-planned
pyre and fire but a gamble nonetheless and one that was incredibly successful. On July 18, 1989, 850 million people, counting those who
viewed the event via television coverage and magazine and newspaper
photographs, watched the burning of thirteen tons of ivory, worth over
three million dollars. In *Wildlife Wars*, Richard described the "enormous irony" (2001, p. 3) of his having spent years preserving prehistoric elephant tusks, only to undertake, in his new position, destroying
over two thousand of those from the present. Yet that is exactly what
happened.

With a life that can be compared to a yo-yo in motion, Richard has had, not counting his youthful safari experiences, four careers. In an August 7, 2003, interview on National Public Radio (http:// www.npr .org/programs/re/archivesdate/2003/aug/leakey/index.html), Richard said the four have "all been fun" and "absolutely spectacular and important and rewarding." Looking for fossils at Turkana was "exhilarating." Building the modern National Museums of Kenya was "very exciting." By 1989, Richard felt ready for career number three when President Moi named him to manage the corrupt and bankrupt Department of Wildlife and Conservation Management, which he reorganized into the Kenya Wildlife Service (KWS). Richard knew he would miss working with Meave in the field, as they had for twenty years, but the opportunity was impossible to refuse. He eventually called those five years directing KWS, with dual goals to stop poaching and establish a wildlife authority, "very, very challenging indeed." Which really was an understatement.

Although he respected his predecessor, Richard inherited a "mess" (Leakey & Morell, 2001, p. 31). His office, stacked to the ceiling with dusty files needing immediate attention, symbolized both the state of the organization and a problem that had been growing for many years. In his father's youth, huge herds of elephants, giraffes, zebras, even flocks of ostriches, blanketed the vast plains. The privileged aristocrats of the 1900s used Kenya's wild game as a shooting range and indiscriminately decimated herds and flocks in "unmanaged slaughter" (Leakey & Morell, 2001, p. 29). Finally, in 1945, a national park service was established. The parks were self-sustaining until 1976 when they became a government agency and received far less money from the general coffers than they took in. With no funding to pay personnel and buy equipment, poaching and corruption became rampant. Tourism helps drive the Kenyan economy, and tourists go to Kenya to see wildlife. Elephant is one of the "Big Five," along with lion, African buffalo, leopard, and rhino,[1] that visitors want to see, but poaching had reached astronomical proportions and gangs were also killing visitors and park rangers in their lust for ivory. On one of the first days on the job Richard wrote in his journal, "No money, no morale, no vehicles, planes are grounded, inadequate senior staff, no fuel, no ideas" (Leakey

& Morell, 2001, p. 48). Nor did he know anything about elephants, conservation, or even how to wear the hat that went with his uniform.

Richard responded exactly as he had when his life was threatened by kidney disease. He allowed no one to know how desperate he felt. The talents that made him successful in fossil finding—organizing and leading a team, fund-raising, and capturing public imagination—were strengths, but to exploit them he needed a plan. This plan had to send a message that reached three different audiences. He had to send poachers a message he was serious about stopping elephant poaching and all trading in ivory; send wildlife employees a message that would improve morale; and send the world a message Kenya was a safe place to see wildlife, but that this vacation oasis would be ruined unless the public stopped buying ivory trinkets and jewelry.

The plan came to him in the guise of a newspaper article criticizing him if he allowed the sale of high-quality ivory in a rigged deal at a vastly lower price than it was worth. Instead of selling the tusks to add much needed money to Kenya's coffers, Richard decided to burn them. Sending the equivalent of three million dollars up in flames would send a fiery message to the world.

Unfortunately, ivory does not burn. Elephant tusks are actually incisor teeth of hard dentine that contains calcium. From his work with fossils, Richard knew teeth are the most difficult part of a body to destroy; a crumbling prehistoric jawbone could have teeth in excellent condition. He met Robin Hollister, a friend's friend who worked with Hollywood special effects and who promised him the "huge, spectacular fire" (Leakey & Morell, 2001, p. 91) Richard wanted for maximum world attention. The tusks would be painted with a clear plastic that readily burned and stacked around a core of straw and firewood. Gas and kerosene would then be pumped into the bales of straw. What Richard didn't want was a newspaper headline blaming him for torching the president of Kenya. With Hollister's help, the world gasped as the huge pyre of tusks ignited and burned. The fire was a magnificent success. Together with elevating elephants on a list of animals threatened with extinction (the Convention on International Trade in Endangered Species, or CITES, list administered by the United Nations secretariat), an ad campaign, and publicity in newspapers and magazines, the price of ivory soon fell from one hundred dollars to five dollars per pound.

Richard freely admits he backed into the job of conservation simply to stop the poaching of elephants. Poaching elephants sounds rather mild, a crime as soft as a poached egg, one that warrants only a slap on the wrist. It doesn't send a visual image of these magnificent, intelligent creatures, each with a distinctive personality, head erect, ears flapping, agile despite size, possibly with a small baby playfully tagging beside. Poachers, usually armed with assault rifles, travel in small groups, but elephants soon become aware of death and communicate about the approaching danger by rumbles that warn others in their family who may have wandered off. The poachers want only the tusks of the elephants and leave behind the bodies, "bloated gray carcasses . . . like rounded boulders . . . scattered across the landscape, some adults, others juveniles: probably a family group trying to run away from their tormentors . . . very still and silent and ineffably sad . . ." (Leakey & Morell, 2001, pp. 59–60). Elephants are one of the most socially aware of all animals and are traumatized by what happens to others in their herd. When elephants return to find the dead, they grieve like members of a human family. "There are frequent reports of elephants picking up and fondling the bones of dead family members, touching their remains gently with their trunks, or covering dead elephants with branches" (Leakey & Morell, 2001, p. 222).

Richard had many reasons for taking on the job of director of KWS, but putting a halt to the poaching of elephants was a major one and he worked hard and enthusiastically. He rid KWS of 1,640 employees who were corrupt and personally profited from the trade in tusks or employees he felt did not earn their salary. Wildlife employees were demoralized by poor pay and lack of equipment to do their job, so Richard tapped many sources, public and private, for funding. Vehicles and equipment, like night-vision binoculars and gun sights, helped KWS track the poachers, who often buried tusks and returned later when they were less likely to be spotted. Success sent the message to others about the heavy price they would pay for poaching. He built staff morale by personally visiting even the most remote parks. He also worked to develop a positive relationship with the World Bank, an organization he had frequently criticized because it gave money to governments without insisting they be accountable for how it was spent. He even investigated working with a researcher in reproductive physiology at a zoo in Cincinnati, Ohio, about a "frozen zoo," a project that involved taking embryos fertilized in vitro from captive parents and trying to put them into a wild mother (Eisner, 1991).

When Richard became director of KWS in 1989, herds across Kenya numbered fewer than 20,000 elephants and poachers were robbing Kenya of about three elephants a day (Leakey & Morell, 2001, p. 36). By 1993 poachers were claiming thirty elephants for the entire year (Leakey & Morell, 2001, p. 278). But success brought about another problem. KWS was responsible for all wildlife in Kenya. Along with protecting game inside their parks, KWS had to protect people and domestic animals from elephants, buffalo, lions, and other potentially life-threatening wildlife that roam the Kenyan countryside. KWS built various types of electric fences that used solar power to keep animals inside park reserves. Tall fences could be dangerous to large animals as they migrate over the vast savannas of Kenya and neighboring Tanzania and the intelligent elephants often figured out how to circumvent the fences, even putting branches on wires to cause a short circuit. One of the more effective fences rose only eighteen inches off the ground so that the elephants received a jolt at their knees but small animals scooted underneath.

Solving each problem seemed to spawn a new one. On September 8, 1991, Richard, Meave, and friends who were members of other conservation organizations watched about one hundred elephants cavorting at a river in one of the parks. Older elephants bathed while the baby elephants ran and played, wrestled with each other and squirted water. Above, planes flew low over assigned blocks of land and elephant counters watched to see if the gray on the ground below wiggled and was an elephant or simply a cloud's shadow. The best news was that the number of elephants at Tsavo East National Park had increased from fewer than 5,000 to 6,500 (Leakey & Morell, 2001, p. 214).

The very next day, before Richard could let the newspapers know that KWS was winning the battle against poachers, he received word his enemies had convinced President Moi that the 1,640 redundant employees had to be reinstated. Richard fought back but each battle took time and energy. During his early days on the job, President Moi warned him to be wary of his safety, and threats on his life required around-the-clock bodyguards for Richard, his home, and his family. He said little to his mother, but she well knew the danger with which he lived and he appreciated her unexpectedly telling him that his work in saving elephants was much more important than finding fossils (Leakey & Morell, 2001, pp. 136–37).

## THE CHALLENGE OF A CHALLENGE

Richard wrote in *One Life*, "As a family the Leakeys have always relished challenges" (p. 21). Thankfully so, because few people have faced more challenges than Richard.

June 2, 1993: To save time, Richard decided to pilot four staff members to a meeting. The plane was in the air only ten minutes when he lost engine power. He was able to glide the plane away from buildings and land on a rough field but one wing clipped a tree. Although the other passengers were not seriously hurt, Richard had to be transferred to a Nairobi hospital. Meave had just set up a field camp at Koobi Fora but quickly turned responsibility over to twenty-one-year-old Louise and left to stay with her husband day and night. If the crash had not been an accident, an assassination attempt in the hospital might follow. Guards were posted outside his room and, despite Meave's anxieties, Richard directed KWS from his hospital bed.

Besides his crushed legs, doctors were worried about an infection that could cause kidney failure. By day five his doctor insisted he be treated in England, where facilities were better. Meave and Mary, two very pushy women according to Richard (Leakey & Morell, 2001, p. 260), agreed and so, fortified by wine allowed by the doctor and dispensed by Meave, he suffered a long, miserable plane trip.

Just as with his kidney surgery, the family's dark humor provided an outlet for their concern. His left leg required amputation eight inches below the knee, which prompted daughter Anna to suggest that the book he was planning, which eventually turned into *Wildlife Wars*, be titled *One Foot in the Grave*. Richard enjoyed his best laugh in many days. Richard himself asked for the amputated leg so he could bury it at Koobi Fora, the only part of his funeral he'd be able to attend (Leakey & Morell, 2001, p. 262). And in a 1999 interview when

Richard was asked if he would like to be cloned, he replied, "No, but it would be quite useful to have several spare parts on hand. Kidneys and some new feet would be of immediate appeal" (*Discover*, 1999).

Richard suffered through fourteen operations to save his right leg and then decided that he needed to return home, clear his mind, and consider if saving the leg were worth the price of more operations. He also fretted about potential future problems similar to the painful osteoarthritis his father suffered. During the next few weeks he thought about his options. Back at the hospital in England, rather than have reconstructive surgery, he elected to have his right leg amputated and both legs fitted with prostheses. He was soon walking tentatively with canes, then quickly learned to navigate on his artificial legs.

Richard credited the crash with bringing him and Meave even closer. He was determined to live as normally as possible. Treating his artificial legs as extensions of his car's brake and gas pedals put him on the road again (Leakey & Morell, 2001, p. 277). Despite his positive attitude, the loss of both legs was devastating to someone with an active life that included trekking over the desert and searching for fossils. He told Virginia Morell, "I lost so much body surface that I no longer shed the heat efficiently, and I break out in these terrible rashes" (as cited by Morell, 1996, p. 32).

Complicating the situation was the possibility that the engine failure was engineered. He had begun his book, *Origins Reconsidered*, written with Roger Lewin with an original publication date before the crash, by stating that "I never quite forget that there are people who would rather see me dead than alive" (Leakey & Lewin, 1993, p. xiii). Although investigation never established a cause for the crash, Richard believed it was not an accident. He has always had supreme determination and self-assurance that make him appear arrogant to many people (Leakey & Morell, 2001, p. 277) and, given his high-profile position directly responsible only to President Moi, his well-publicized success in stopping the trade in ivory, and his long history in standing up for causes in which he believes, there's likelihood that someone tampered with the plane's engine.

His nameless political enemies didn't stop their work during Richard's recuperation. On March 23, 1994, less than ten months after his plane crash, Richard, angry and frustrated with moves attacking

KWS's staff and undermining gains made to protect the parks and the animals living there, resigned.

Two days later biologist David Western was appointed director of KWS. Richard admired Western's integrity but they had a history of disagreeing over conservation policies and he felt Western had long been angling for the position. Western began a program that focused not on the parks but on the communities that lived near the parks. Richard always tried to hire people from the community in which a camp or park was located, to assist with medicine and food needs, and support children in obtaining a high school education, which is not free in Kenya. He adamantly opposed simply giving money to people because it "runs the very real risk of turning villagers into welfare recipients" (Leakey & Morell, 2001, p. 206). The role of local communities in conservation efforts is a contentious one in a country in which wildlife is entwined with village life, but Richard has always maintained that national parks are "the last safe haven for Kenya's wildlife" (Leakey & Morell, 2001, p. 280).

In addition to disagreement over conservation philosophy, Richard realized Western lacked experience in two key areas: managing a large team of employees and standing up to government politicians. At KWS, Western would have a great deal of dung thrown at him and it would be a lot harder to dodge than the real stuff at Laetoli!

## RICHARD LEAKEY, POLITICIAN

When Kenya, with its multiple tribes, gained independence from Great Britain, a one-party system seemed logically able to reduce intertribal rivalry and fighting because the president could balance tribal participation in the government. By the late 1980s relief from being rid of colonialism gave way to desire for a multiparty system. Kenyans who wanted a more open political structure were aided by pressure from the State Department of the United States, through Ambassador Smith Hempstone, who actively urged a two-party system. Although Moi was upset by the meddling in Kenya's affairs, illegal political parties had already begun to form. As director of the Kenya Wildlife Service, Richard had to stifle his interest in politics, but he wrote in his diary about yearning to be part of Kenya's political future and questioning why President Moi tolerated corruption and other abuses. Richard realized a multiparty system was necessary if Kenya were to mature and grow. By 1992 pressure outside and inside Kenya opened the political system, but Moi was elected to his fourth term as president amid reports of vote tampering and not allowing certain segments of the population to vote.

Leaving KWS's directorship provided the opportunity Richard needed to pursue politics and energized him after the ordeal of losing his legs. He felt sure a coalition of the smaller political parties could unseat Moi's Kenya African National Union (KANU) party. Safina, Kiswahili for *ark*, was formed in 1995, and the party leaders, including Richard, who was one of its founders and Safina's Secretary-General, vowed to battle corruption and human rights abuses. Kenyan politics again treated Richard and his colleagues roughly. Richard was accused of being a racist, a traitor, and a terrorist. In one sad episode he and other members of Safina were attacked by a mob of young men armed with whips. Richard, at a definite disadvantage with artificial legs, managed to escape with only broken car windows and felt lucky to be alive (Leakey & Morell, 2001, p. 297). Little wonder that he and Philip, who led a group of eighty-seven white Kenyans in support of Moi and the KANU party, were again at odds with each other. Even though Safina was deprived of registration as a legitimate party until two months before the December 1997 election, members won five parliamentary seats. Richard had not sought one of these but was nominated and officially became a member of Parliament.

Losing his legs gave Richard a deep appreciation for the problems the average citizen with disabilities encounters, especially in a developing country like Kenya. He suffered difficulties and embarrassment, most notably when he first met with President Moi on his return from amputation of his right leg. Eager to present his best face to the president, he instead landed on his face. Luckily a photographer was not present. He vowed to work for a better life for Kenyans with physical disabilities, many of whom lack money to purchase expensive equipment that would help compensate for their disability. Never one to miss an opportunity, Richard entered Parliament to take the oath of office not walking on his high-tech artificial legs but in a wheelchair. With no ramps, his staff had trouble helping him manipulate through the building. As Richard hoped, the problems facing Kenyans with physical disabilities made front-page news.

## AND BACK TO CONSERVATION . . .

A year later, 1998, President Moi asked Richard to return to the Kenya Wildlife Service, conveniently ignoring that just a few years before he had considered Richard "a sworn enemy" and called him "a racist, a colonialist and a foreigner" (Harden, 2000). The agency again had major money problems. Meave was supportive although amazed that Richard was willing to return (Leakey & Morell, 2001, p. 302). He resigned from Parliament, but not from Safina, and for the first time sat in the director's office of the building he had designed and whose construction he had overseen.

## AND BACK TO POLITICS . . .

Richard remained with KWS only one year before Moi asked him to take another government position, this time as head of a "Reform Team." In describing this fourth job on National Public Radio, Richard said, "Wow!" and rhetorically asked how many people in a lifetime get a chance to head up an agency of 500,000 people working to stop corruption and bring economic reform to their country (R. Leakey, 2003). Less than two years later, with Richard beginning to make headway in the agency's efforts to stop corruption and bring economic reform to Kenya, Moi again fired him.

## AND BACK TO CONSERVATION . . .

Although Richard had been deeply involved in several different conservation efforts since he first took control of KWS, he steadfastly held to his basic philosophy that money from tourists, most of whom come from the Western world, should pay for protecting African wildlife. An alternative route is to kill the animals and garner money from sale of meat and tusks. Richard never believed animals had to be sacrificed, nor does he believe that Kenya and other developing countries of Africa should "shoulder alone the expense of preserving its wildlife and wilderness" (Leakey & Morell, 2001, p. 224). Feeling strongly that the wealthier nations of the world must financially support the poorer nations in efforts to preserve threatened game populations, Richard believes, "The world must wake up to the fact that poor countries can't bear the financial burden and arrest their development simply because the richer countries feel sentimental" ("Scout Report," 2004). In other words, if people in wealthier countries are upset with the thought of poachers killing elephants or of apes starving because their habitat is taken, they must, to use an old cliché, put their money where their mouth is and contribute to programs to protect wildlife.

Today, Richard is a spokesperson for the Great Apes Survival Project (GRASP) sponsored by the United Nations, which hopes to raise $25 million over the next three years. "There's a lot of waffle about how the apes are our closest relatives, and so on, but while we talk, they go," he says, emphasizing that all the great apes face possible extinction within the next half-century without help.

As a visiting professor at Stony Brook University in New York, Richard continues to write books that enjoy public popularity. His 1995 book, *The Sixth Extinction*, written with Roger Lewin, deals with another mass extinction they believe the world faces. This one, they write, won't be caused by climatic catastrophe but will be the result of human failure to protect earth's species. Part of Richard's writing appeal in all of his books is that, as British reviewer Tim Halliday wrote, Richard can "step back and set his own findings within the broader context" (Halliday, 1996).

## MARY AFTER OLDUVAI

Mary continued to live in Nairobi until her death December 9, 1996, at age eighty-three. She didn't garner the public renown afforded Louis, but she did earn respect and admiration from many of her colleagues. Although Louis was "less than meticulous" (Lewin, 1997, p. 133), a statement Rick Potts of the Smithsonian Institution supports, Mary was painstakingly particular and set new standards for excavation. Rick Potts calls Olorgesailie Louis's project and says that the only site there accurately plotted and carefully excavated was the one on which Mary worked (Potts, personal communication, October 26, 2004).

One evening when Rick Potts and Mary were enjoying a drink at the end of a workday, Mary confided that a scientist is fortunate if an idea of his or hers outlives the scientist. Fossil finds, however, live on. He knew she was specifically referring to a recent article in which he refuted something she had written earlier, but he also recognized that she was referring generally to her legacy of specimens. *Proconsul*, Zinj, a multitude of tools and fossils found at Olorgesailie and Olduvai, the trail of footprints at Laetoli—they were the legacy Mary left for future generations who could then interpret them according to their own perceptions (Potts, personal communication, October 26, 2004).

Despite the wealth of fossils and the trail of footprints she left behind, perhaps Mary's most important legacy was encouraging granddaughter Louise to build on the family's heritage.

## LOUISE

Louise Leakey says she became a paleoanthropologist "by default" (Eddings, 2003) when her father's plane crashed. Nasser Malit, who began fieldwork with the Leakeys in 1993 while still an undergraduate at the University of Nairobi, believes both this event and, three years later, the death of her grandmother Mary influenced Louise's career. "Word was going round that she was not as enthusiastic in fieldwork, but her grandmother kept encouraging her. She was very close to her grandmother, as I understand. She stayed close to her till the time of her death and subsequent spraying of her ashes over Lake Natron in northern Tanzania" (N. Malit, personal communication, June 1, 2004).

Like her father, Louise learned as a young child how to find and extract fossils from the field, but she was unsure if she wanted to spend her life in the scorching sun. Nasser Malit remembers her saying she preferred a cool drink in the shade (N. Malit, personal communication, June 1, 2004).

Organizing and running a camp in a remote desert area is a difficult job that even today requires immense organizational responsibilities. Although she was only twenty-one and beginning her college studies in England, Louise filled the void when Meave left quickly after Richard's plane crash. Even though Louise had much experience in the field, Malit feels the 1993 season "had the greatest influence on her after being left with the sole responsibility of taking care of the camp and day to day running of fieldwork activities" (N. Malit, personal communication, June 1, 2004). She survived with support from the more experienced leader of the "Hominid Gang," Kamoya Kimeu, but running the camp at Koobi Fora was difficult for Louise, and crewmembers preferred to deal with Meave. Ten years later in talking with both crew and researchers, Nasser Malit "only heard good things about her" (N. Malit, personal communication, June 1, 2004).

## MORE CONTROVERSY

The Leakeys generate extremely strong feelings in the world of paleoanthropology. Although they have many supporters, other scientists feel popularizing their work with the public cheapens it and resent the press the Leakeys have consistently received since Louis's early fieldwork. Others seem to begrudge the huge financial backing of the National Geographic Society, which they see as a pseudo-scientific society rather than a "pure" scientific institution. Some folks just enjoy a chance to challenge the family. And others disagree with them on issues related to paleoanthropology.

Not to be left out of the family tradition for controversy, Louise joined Meave by throwing her hand into the mix as well. In 2001 when a team headed by both of them announced *Kenyanthropus platyops*, or "flat-faced man of Kenya," they set off, to little surprise, another paleo-maelstrom.

*K. platyops* has a unique mixture of features, with a skull similar to *A. afarensis* and *A. africanus* but with a very flat face. Its age is about 3.5 million years, the same age as Lucy. Again the question: Did multiple species live at the same time?

When Ray Suarez, Public Broadcasting System (PBS) commentator, interviewed Meave Leakey, he asked, "Is it a pretty audacious thing to make this an entirely different separate genus and species really separated out from the fossil record so far?" Meave replied that audacious wasn't the correct word. "I think it's quite bold. But we wanted to show that this is unlike any other genus known at the time" (Online News Hour: New Beginnings, March 26, 2001).

If *K. platyops* existed almost four million years ago, then Lucy, or *Australopithecus afarensis*, has another contender for first human ancestor. Many in the scientific world, like Daniel Lieberman, are unsure. He wrote in *Nature* magazine that *Kenyanthropus platyops* raises "all kinds of questions, about human evolution in general and the behaviour of this species in particular" (March 22, 2001, p. 420). Lieberman believes the new challenge will be for skeletal biologists, paleontologists, and molecular biologists to develop new techniques for obtaining data. His best guess is that it will be a long time before the scientists know where to place *Kenyanthropus platyops* (KNM-WT 40000) on the human evolutionary tree (p. 420).

Tim White has no doubts. He believes this new Leakey find should carry Lucy's name, *A. afarensis*, because geology, not genes, caused the distinctive flat face. White theorizes that soil as fine as sand settled into tiny cracks in the skull and, over time, resulted in a misshapen head. Although White admitted to viewing Flat-faced Man but not studying it, he recalled pig skulls in which geological contortion made analyzing the fossils difficult (White, 2003, p. 1997).

At the heart of this contention is a philosophical argument about how best to categorize human ancestors. Tim White, a lumper who believes in fitting hominin fossils into the fewest possible species, attributes the different characteristics of many fossils to variations on a theme. If someone looks at a tall, stocky basketball player and a short, slightly built jockey, the differences between the two are immediately seen and obvious. If the bones of each are found in several million years, will they look as if they came from the same species? A paleoanthropologist struggles with this problem every time he or she confronts a new hominin fossil. If the person is a lumper like White, the tendency is to place the fossil into an existing category. White's family tree shows a linear relationship, with species leading out from the tree's trunk (DeSilva, 2004, p. 259), but with dotted lines to show relationships are not absolute.

Meave Leakey, on the other hand, believes several species of humans, like other animals, existed at the same time, with some becoming extinct. Instead of a time-honored family tree, she prefers to show relationships by organizing species into groups within shaded ellipses, such as *Paranthropus aethiopicus* (KNM-WT 17000) and *P. robustus* (SK 48 from Swartkrans, South Africa) and *P. boisei* (KNM-ER 406 and OH 5, from Olduvai) (DeSilva, 2004, p. 263).

In the *Human Evolution Cookbook* (Dibble, Williamson, & Evans, 2003), splitters and lumpers are seen as two competing teams. The splitters, as the game is described, find fossil bones and give each microscopic piece a name, one of which is the "missing link" paleoanthropologists have been searching for since Darwin. The announcement brings this team much attention.

> Anyway, after allowing the splitter some time in the spotlight, the intrepid lumper goes forth and declares that half of the scraps found by the splitter actually represent the same species (sometimes even the same individual), and the other half are just pieces of fossilized chickens. Ergo (Latin for "ha, ha"), the new finds should all be given the same name, and more important, they should be given the name of the first-recognized member of that species (who was, most often, found by the lumper's uncle). This makes the splitter very angry and for a couple of years there are nasty exchanges as hundred of acres of prime forest are turned to paper on which numerous articles and books are written, published, and then forgotten. (Dibble, Williamson, & Evans, 2003, p. 20)

Then the game starts over.

Unfortunately, this humorous view of splitters and lumpers has a serious side in a field that "elicits more passion, skepticism, and debate" (DeSilva, 2004, p. 257) than any other area of science. After an article in *Science* in which he built a case that KNM-WT 40000 was not a new species, White shot at Meave and Louise when he concluded, "Past hominid diversity should be established by the canons of modern biology, not by a populist zeal for diversity" (2003, p. 1997). Louise has said only time will tell if *K. platyops* is a new species (personal communication, October 10, 2003).

The Publisher's Prologue to Louis Leakey's posthumous autobiography, *By the Evidence: Memoirs, 1932–1951*, compares the study of human evolution to the game of croquet in *Alice's Adventures in Wonderland*, where the wickets walk away. "Fossils cannot walk away, but they do seem mysteriously to change characteristics when viewed by different observers (p. 8)." A new species or even genus to one set of trained eyes is an insignificant variation to another, and again, the issue rests on whether one is a lumper or splitter. The Leakeys have given names to some fossils that didn't deserve a separate species and have correctly assigned names to others. Regardless, the family has brilliantly produced fossils that have starred in the human evolution drama for almost one hundred years.

## NOTE

1. The term "Big Five" comes from the days when hunters confronted these animals with guns, with the goal to take trophies home. Poachers have hunted the African rhino, both black and white (named not for color but for its wide mouth), almost to extinction. Hunting is now outlawed in Kenya although only controlled in nearby Tanzania.

# CHAPTER 10

# BUT WAIT, THERE'S MORE...

## *(2004 and Beyond)*

As philosophers constantly tell us, and we keep finding out the hard way, science is tentative, existing perceptions are constantly replaced by new ones.

—R. Leakey & R. Lewin, *Origins Reconsidered,*
Prologue, p. xviii

## HOW HUMANS CAME TO BE

Who are the first human ancestors—those from which all humans today are descended? How humans came to be is a mystery equal to any fiction found on television or movie screen. Think of the world's prehistory as a drama—and it is—one that divides experts and tantalizes the audience. For years, it was believed there was a single main actor from whom all the earth's people are descended— literally a one-person show. Currently, however, evolutionary evidence has proved that many parts are played by a variety of actors and no one yet knows who plays what part. Instead of a Terminator who returns to change the past, somewhere, probably buried in African soil, there is an Originator whose family changed the present. Who is the Originator of this human drama? Will Lucy play a major or minor role? What role will her cousins, who left their footprints in the ash at Laetoli, have? Is

*Homo habilis* the closest ancestor to *Homo erectus?* Will 1470's brain make him a major player? Does *Kenyanthropus platyops* deserve a curtain call? Or is *K. platyops* the last of his line, with a very small part, or perhaps related to Lucy and her family? Who will be listed in the credits as the star with the role of Originator? Who will be listed as the producer who found the star? What twists in plot await? And last, since this is a play with a great deal of suspense, how long must the audience wait to find answers to the crucial questions: How did humans come to be? And when did it happen?

In the last one hundred years much has become clear about human origins but much is yet to be learned. Fossils have recently been found in Chad, Kenya, and Ethiopia that support Louis Leakey's prediction that the ancestors of early humans appeared far back in time. Though not so fast-paced as an action movie, the plot of this drama is still unfolding and the ending is impossible to know. But the chase for the Originator does matter.

Today Louise Leakey, just like her grandfather, firmly believes getting more fossil evidence is necessary for finding answers to the critical questions about how humans came to be. Understandably, people who have daily needs for survival often do not realize the importance of investing in human origins research, but Louise feels people must come together and recognize a shared common ancestor in this time of so much tribalism, ethnic differences, and human conflict (L. Leakey, personal communication, October 16, 2003).

Finding evidence of these early human ancestors is a complicated task. The headlines announcing a major fossil find don't detail the long, grueling process required to do the job. The headline on page A-1 of the *New York Times* on March 22, 2001, read "Skull May Alter Experts' View of Human Descent's Branches," but the article neglected the story behind the story. It ignored the preplanning and organization needed to find *Kenyanthropos platyops*.

Although improved by today's technology and equipment, fieldwork is still difficult and exacting. Apart from fund-raising, organization of crew and equipment, and managing camp life, working in sand that radiates one-hundred-degree heat and using a dental pick and small artist's brush to dislodge a multimillion-year-old bone from its desert grave is not easy work. However, over the years the legendary Leakeys have found excitement doing exactly that. As Richard wrote, "the sheer magic of the enterprise drew me into it" (R. Leakey, 1994, p. xi).

## SETTING UP CAMP

A well-equipped, smoothly functioning camp is a very costly affair in terms of money and organization. Louise Leakey still has to constantly confront the challenge faced by her grandparents and parents of finding funding. Many people assume the Leakey Foundation, established in 1968 by Californians Helen and Allen O'Brien, automatically provides the much-needed research funds, but not so. Because patriarch Louis Leakey had broad interests in fields like archaeology, primate research, and ornithology in addition to paleoanthropology, the Leakey Foundation provides grants in a variety of areas. Louise and Meave have to stand in line along with other researchers and organizations that apply. The National Geographic Society is a consistent sponsor and supporter, but other sources must be tapped. Meave travels to lecture about their work and the need for funding and Louise regularly communicates her message while still in the field with online dispatches at http://www.kfrp.com.

Louise's goal is funding that will allow a crew to work full time, year round, rather than for a three-month sojourn, a project estimated to cost $2.4 million over five years. Many of her crew are not very experienced; most are young and need training. Retaining crew members is a relatively new problem because in the past, jobs at the National Museums kept them employed after the field season. If they work in the field for three months but then have no means of supporting their families, they often are lost to other jobs. This revolving door of experienced staff leaving, replaced by new who require training, creates a major problem.

## FROM FINDING FUNDS TO FINDING FOSSILS

Once funds are available, the painstaking preparations for fieldwork begin. Repair and servicing vehicles is crucial. With the rough terrain at

Koobi Fora, vehicles in good condition are a necessity. Tents must also be mended or new ones obtained. Louise has to contact all the workers and let them know the dates of the field season. Meave and Louise also have to work out the schedules and accommodations for other researchers who want to visit. A huge variety of gear and supplies must be assembled, including Global Positioning System (GPS) equipment and aerial photographs, digging tools, small brushes, tents, drums of fuel, and basic food and supplies for over twenty people for three months. At least a day or two may be devoted to packing. Because of weight, as much as possible is carried by truck, a trip of almost four days from Nairobi.

During the field season, Louise herself spends an incredible amount of time flying perishable food and supplies, a five-hour round-trip by air. She recently purchased a previously owned Cessna 206 that can carry heavier loads and is better able to handle short airstrips. However, she lacked the money for extra fuel tanks that would allow her to make the round-trip from Nairobi to Koobi Fora without refueling.

## JUST DESERT

Koobi Fora today encompasses approximately one thousand square miles of raw desert land in northern Kenya's Sibiloi National Park, described by *Travel News & Lifestyle* (Barsby, March 2004), a magazine touting East Africa as the country's "most remote, rarely visited yet world-renowned" park, whose weather is "scorching hot and arid" (p. 24), with very strong winds and an annual rainfall of mere drops. This constant whipping wind that works to uncover fossils and create erosion in sandy areas often turns up specimens on ground well tromped and searched in the past. The Leakeys have mined for fossils in this remote location for almost forty years, but Louise feels the Koobi Fora Research Project (KFRP) still has much work to do on the eastern side of Lake Turkana, "In fact," she wrote at the end of the 2004 fieldwork season there, "we have barely begun" (L. Leakey, 2004f).

Koobi Fora has no shops or theaters, limited electricity, few tourists, some very bumpy roads, and a few four-wheel drive vehicles abundantly in need of repair. Accommodations include picturesque stone buildings with thatched roofs called bandas, and, although they seem primitive compared to a luxury hotel, there are amenities undreamed of by earlier paleontologists. Today an on-site desalination plant processes Lake Turkana's alkaline water to make it drinkable. The main banda, with solar-powered electricity, has a food freezer, a locked area for expensive equipment and valuable fossils, and space for dining. Louise's private banda is not truly private since it is home to a work area and an open-air "great room." It also has a large freezer, and when fresh chicken for visiting National Geographic Society photographers was left behind at Nairobi's Wilson airport, Louise was seen peering into her freezer, like any housewife unexpectedly needing something for dinner guests. A quick run to a grocery store is not an option at Koobi Fora!

Even with improved facilities, there are always unanticipated challenges. When the team of over twenty fossil hunters, drivers, mechanics, and kitchen crew, headed by Meave and Louise, arrived in Koobi Fora in February 2004, unexpected and unusual rainfall flooded the tents. The truck trying to move them to higher ground became mired in mud and several hot hours were spent freeing its wheels and moving the camp.

There is constant negotiation between local people and Louise. Louise has had to change plans when gaining community support was not possible, although Richard helps in negotiating with the local Kenyans. Like his father, Richard has worked for closer ties with his indigenous neighbors. In 1973, Richard requested Kenya's president, Jomo Kenyatta, to grant the Dasanaach people Kenyan citizenship in return for turning in their weapons.

"Thus Richard Leakey has a long history with the people of the area and is a well-respected elder" (L. Leakey, 2004b). Herds of cattle also tromp over Koobi Fora, even though the camp is located in a national park where livestock are not permitted to forage. In an area where grazing land is hard to come by, helping herdsmen understand the damage flocks do when they walk over fragile fossils is an ongoing effort.

## HI-HO, HI-HO

At last, after establishing camp, the real work begins. Crew members have an early morning cup of tea and head from camp a little after six, in early morning light but with the still present coolness and calmness of night. The evening before, lunches, equipment, and a generous supply of water are packed in the truck that carries the crew to the site of the day's work.

After a bumpy ride, the search team walks slowly across the landscape to locate fossils of all types. Dr. Nina Jablonski, chairperson of the Department of Anthropology at California Academy of Sciences, describes successful fossil finding as taking "a lot of knowledge, footwork, patience, and some luck." (Jablonski, 2004) High tech equipment also plays a big part. Geographic Information Systems, or GIS, show a pattern of previous fossils identified by GPS. The maps generated combine with aerial maps, photographs, and ground surveys of an area to decide where the search team will look. During the 2005 season the fossil search team added hand-held field computers to its equipment arsenal. In the past data had to be entered in a laptop each evening; now downloading data through a docking station saves a great deal of time.

Although human fossils are obviously most desired, the team wants to build a picture of the total environment, so bones and teeth of antelope, pigs and other animals are also collected and recorded. Each bit of information adds a piece to the prehistory jigsaw puzzle. In her Week 5 online dispatch from the field in 2004, Louise wrote, "Without the fossils we really cannot convincingly work out the relationships between the different hominids and the palaeoenvironment and habitat in which they evolved." In a three-month period, 500 to 600 specimens are usually collected; typically five to ten of those are hominin.

Once a fossil is found, a small pile of rocks, or cairn, is built to flag the fossil for further examination and a team with higher-tech equipment records a Global Positioning System (GPS) point (or coordinates) and takes digital pictures. With the navigation feature, GPS "is efficient and precise enough to enable easy relocation" (Jablonski, 2004). In addition to fossils, the 2004 project also made an effort to find numbered aluminum tags or concrete bricks marking the positions of hominin fossils found in the 1970s expeditions. A GPS fix quickly updates the database with these old fossil locations.

## THE THREE C'S: COLLECTING, CATALOGING, AND CLEANING

With year-round funding, the need to prioritize the collection of fossils would be minimized, but for now, each evening images are analyzed and decisions made about which fossils to collect the following day. The collection team has its own GPS instrument to locate the fossils, plus a digital camera, aerial map, and field book. Nasser Malit refers to all of this as "cool stuff" (personal communication, October 24, 2004). In the field book information is recorded manually that identifies element, species, geology, and GPS reading as a backup. A picture showing the cairn and the small flag is taken from several meters away to identify the geology. Then a close-up for fossil identification is taken to support identification. As a backup, the collection team marks the fossil's position on an aerial photo with a pinprick. This system was used for many years, but two problems were accuracy and difficulty managing the information over a number of years. Imagine the thousands of pinpricks on hundreds of aerial maps collected over three decades of digging at Koobi Fora! High tech has dramatically altered the old process and allows positive identification and a quick return to the location of fossils the team found previously.

Sophisticated equipment may start the job of collection, but then each fossil is carefully wrapped in common toilet tissue and placed securely in a plastic bag for its trip back to the base camp. For larger fossils that are partially buried, a hole much larger than the specimen must be excavated. A hardener called Bedacryl is applied on the exposed part of a fragile fossil before dental picks and artist's brushes are used to remove excess soil.

If other parts of a specimen might be located in the vicinity, searching continues, especially if the fossil is hominin, even though the likelihood of finding additional pieces is small. In describing the necessity of following up with a tooth, Louise wrote in a 2004 dispatch, "Sadly the remainder of this specimen is likely to have washed away a long time ago but we will have to do a hill crawl here to make quite sure of this." She described a hill crawl as "putting the team in a long line and having them work gradually across the surface turning over every stone and looking for any possible pieces. It is too big an area to screen immediately so we start this way . . ." (L. Leakey, 2004e).

A hill crawl and sieving (English term) or screening (American term) usually happen after lunch. They take time and the chances of turning up anything valuable are slim, but during the 2004 season one crew member found half a "worn and weathered" molar (L. Leakey, 2004b) and most of the other half was found when screening.

At the base camp meticulous care is taken in unwrapping the fossils. Sizzling afternoon heat provides time for each fossil to be carefully laid out in a small box and another digital image made; numbers derived from the field book are used to tag the specimen and are entered into a database. The fossils are carefully rewrapped for storage, with hominins stored in a special wooden box.

An inventory, updated every evening, is made of fossils collected and not collected. By the end of a season several databases exist that back up one another.

## THE THRILL OF THE FIND

Dr. Amman Madan at the Indian Institute of Technology, Kanpur, says, "There is something magical about spending hours poring over every square foot of land and then being able to pick up and hold a part of the ancient past" (personal communication, June 7, 2004). Finding a hominin is the ultimate. In week 11 of her 2004 dispatches from the field, Louise wrote that one of the crew had found an isolated hominid molar "only 50 meters or so from the tortoise house on a flat surface covered in fine wind-blown sand. It is surprising as many visitors must have walked over this specimen for a long time, but as I have pointed out before, unless you are looking for something specifically, fossils and especially hominids are actually quite hard to find!" (L. Leakey, 2004e).

But found they are! In 1999 Patrick Gathogo, now a Kenyan geologist, was with the crew when research assistant Justus Erus spotted a tooth lying in the bare brown soil. Gathogo remembers that the group stopped and leaned over the tooth to get a better view. No one spoke. Gathogo later said, "I knew what they found without them telling me," but very quickly the crew recovered and then "[e]veryone was so excited we started rejoicing" (University of Utah Media Release, 2001). Erus's keen eyes had spotted what is now known as *Kenyanthropus platyops*.

Despite the absence of what most people today generally consider needed amenities, the karma Koobi Fora exudes is a feeling of peace. It offers a fascinating glimpse of time past and present and a sense of the future—seeing a glorious sunset over the jade-green waters of Lake Turkana, gazing into a chasm created endless years ago, watching a raven peck for crumbs from lunch bag remains, visiting an exhibit housing an actual tortoise shell 1.8 million years old and large enough for a baby's cradle. Although a visitor needs guidance to navigate the area, Louise likes the "restful" atmosphere outside Nairobi (L. Leakey, 2004a). She wrote in her 2004 Dispatches about the brightness of the night sky far away from the intrusion of human-made lights, seeing a family of warthogs near camp with two piglets (L. Leakey, 2004d), and the "stunning sunrise in the mornings." (L. Leakey, 2004b). Her father, Richard, wrote in *Origins Reconsidered*, "I love the desert. . . . To some, such terrain seems hostile. To me, it is like coming home, and I feel a sense of peace" (p. 12).

The difficulties of living and working in such a remote area encourage a sense of camaraderie and mutual support. The camp acts as a dispensary and provides drugs to local people needing first aid or basic medicine. A plane flying to Koobi Fora carries additional passengers and supplies. The Koobi Fora Research Project crew helps the Kenya Wildlife Service launch a boat donated to Sibiloi National Park by the International Fund for Animal Welfare. Part of the funds Louise and Meave raise go toward community outreach programs at the nearby village of Ileret, and Louise herself works with a project to renovate a health clinic and train staff and to build an extension to the primary school.

Not to say that events at Koobi Fora don't present challenges, both ongoing and unique. Mechanical difficulties are a constant problem. Although most of the camp's equipment runs on solar power, the water desalination plant requires a generator. Louise calls the electrician who keeps it running "brilliant" (personal communication, March 19, 2003), but getting him to Koobi Fora and back to Nairobi when problems arise is difficult. With rocky roads that are little more than dirt tracks, cars also need constant maintenance.

Other challenges occur rarely. When lecturing, Louise shows a slide of her standing atop her small airplane, hoping the additional height will allow her to contact a hospital in Nairobi. Although the view from the roof of the Cessna stretches dramatically out over the Turkana Basin, Louise is clearly intent on begging for cooperation from the plane's radio she holds in one hand or the satellite phone in the other. Evacuating a ranger with a poisonous snake bite by the plane on which she is standing is crucial, since a motor vehicle trip over bumpy roads would take far too long to save his life. Most of the year the airstrip is dry and, although pockmarked with the hoof prints of wild game, bumpy but usable. On this day the plane's wheels were mired in mud that made take-off impossible. The entire crew worked to dig out the plane's wheels, they managed to take off, and the ranger survived his scary scuffle with death, "but only just," Louise says (personal communication, October 10, 2003).

Eventually the season comes to a close. For the trip to Nairobi's National Museum, especially rare fossils are coated with a plaster jacket. All of them are placed in individual plastic bags and surrounded by packing material. Once stored at the museum, they leave the security of a vault only for careful cleaning and study. By law, to protect these irreplaceable national treasures, no hominins leave Kenya, even for exhibition at other museums. Researchers travel to the museum or work with replicas.

Finding fossils that represent human beginnings has been a mission of the Leakeys through three generations. Today Louise continues that tradition by encouraging indigenous Kenyans to join in this mission. When she talks about finding fossils to piece together the human past, "It's not about Leakey," she says. "It's about the young Kenyans—for them to carry the flag" (personal communication, March 19, 2003).

## NASSER MALIT

Nasser Malit is one of the young, intelligent, and ambitious Kenyans who epitomizes Louise's vision. He was born in 1970 in Nairobi but grew up in a small village near Lake Victoria in Nyanza province where he went to both primary and secondary school. During his secondary school days, Nasser says his biology teacher had a lot of influence on him. His teacher's mentoring paid off when he eventually passed the Kenya Certificate of Secondary Education examination to enter the University of Nairobi.

As an undergraduate majoring in anthropology, Nasser volunteered at the National Museums of Kenya in 1993 and was granted an internship. Meave was soon leaving for fieldwork in Lothagam, on the southwest side of Lake Turkana, and said yes when he asked to join her. Less than two weeks later, Richard's plane crashed and Meave had to leave the field to tend to Richard. During this field season Nasser met Louise, who was called back from college in England to take the leadership role in the field. Besides working with Louise for the first time, Nasser worked with heady company that season, with notable scientists like Alan Walker, Dr. Craig Fiebel, a geologist then with Utah Geology School, and Emma Mbua, then with the National Museums of Kenya and now head of the Division of Paleontology there. Nasser also worked with experienced fossil hunters like Kamoya Kimeu, Wambua Mangao, Peter Nzube, and other young members of the Hominid Gang, as they are popularly known. Nasser now says, "This season turned out to be the turning point in the direction my career would take" (personal communication, June 1, 2004).

Nasser, like other young Kenyans on the team, has benefited over the years from Meave's mentoring and support. Meave has been keen on crew training in areas like fossil preparation, computer skills, and museum organization and management and has encouraged most of the crew to further their education.

After his graduation from college, Nasser was eventually employed in 1996 in the Division of Paleontology at the National Museums of Kenya. In 1999 he was granted a study leave that enabled him to take up graduate studies at State University of New York in Binghamton under the supervision of Prof. Philip Rightmire, a prominent scholar in paleoanthropology circles. Nasser spends his school years there with his wife, Dorothy, and son, Oliver Kinda. His doctoral dissertation focuses on taxonomic and evolutionary relationships among *Homo erectus* populations in Africa and western Asia, especially the recently described fossil hominins from Dmanisi, the Republic of Georgia.

Over the years Nasser has been a close scientific collaborator with Meave and Louise. This led to his being invited to direct and excavate a late Pleistocene hominin skeleton in northern Kenya's Samburu District. It is this project that in the summer of 2004 took him back to Kenya. Even after acquiring research permits from the National Museums of Kenya and relevant Kenyan ministry, Nasser says when he arrived in Samburu to start working on the site he realized the local people weren't happy with his digging "something precious from their land," which is also a protected area under the Kenya wildlife conservation act. "It was," Nasser says, "a very tense and frustrating situation that nearly killed the project." Several more weeks were spent in negotiations with Samburu elders who govern the area, and Nasser had to work with a variety of people in the wildlife and local government ministries and eventually spoke to Samburu Council leaders who bluntly wanted to know how the project would benefit the community. Nasser won the difficult negotiations when he volunteered to make a cast of the specimen. Taking this replica back to the site would help to publicize the area as a tourism spot. Negotiations, he found, took longer than the actual digging! This, he says, is a typical situation for the Leakeys and other paleoanthropologists. Nasser anticipates announcing to the world the existence of yet another very important hominin from eastern Africa, dating close to 100,000 years BCE.

As president of the African student community at Binghamton, Nasser encourages Africans to return home after graduation, as he plans to do. He came to the United States because there was no graduate paleoanthropology program available in Kenya, but one of his dreams is to develop such a program. He would like to team with Louise to be "an institution builder." Ever since his undergraduate days, he and friends have vowed to set up a strong Kenyan paleoanthropology program that would be associated with a university and incorporate related fields, like skeletal biology, gross anatomy, skeletal pathology, and paleoenvironments. He believes the potential is there and it would benefit many people. He says, "I grew up in a family with economic hardships and economics will play a role in my decisions." He adds that "The world has become so global that you cannot ignore any culture" and hopes his training and American sojourn will be valuable to his work back home (personal communication, November 26, 2004).

Like Louise, Nasser is also anxious for a twelve-month research program in Koobi Fora. He says, "The challenge of keeping Koobi Fora camp an active research station running all year long is just one of the greatest challenges in her present life and I share her passions for this kind of development" (personal communication, June 1, 2004).

## LOUISE'S PERSONAL PLANS

Louise was married in December 2003. Her husband, Emmanuel de Merode, does conservation work in the Congo. He has built his own plane powered by a Subaru car motor and flown it across the Atlantic. He still flies it, doing game counts in the protected areas in the Congo, and for the time being is working with the European Commission based out of Goma.

The couple is now building a house complete with vineyard in northern Kenya, where they have their base, and enjoy holidays along the coastal area of Mombasa. Louise and Emmanuel named their first child, born in 2004, Seiyia. Seiyia is the name of a big sand river in northern Kenya. "It's a beautiful name so that's why we used it and we love the north. I am sure Seiyia will grow up feeling the same way, as we will be spending much time there in the coming years," Louise wrote (personal communication, October 12, 2004). Like Louise, Seiyia began treks to the field while a baby and joined the 2005 expedition to Koobi Fora at age five months.

Louise and Meave plan to continue their work at the Turkana Basin so long as it seems to be leading somewhere, but there are complications other than funding. The quarters Louise needs for her work belong to the National Museums of Kenya and are available for public use. Although few tourists choose to visit the Koobi Fora camp, Louise has found strangers on her patio and, worse, handling fossils as if they were flipping through items in a store. Even with a possible alteration in where she is now digging, Louise's plans include "working in the north on the various projects—prehistory, and with the park authorities in trying to raise some support for the long term protection of these fragile and important areas. It's hard to know what we will end up doing but both my husband and I will probably spend some time working in Kenya together in the future" (personal communication, October 12, 2004). Louise sees a bright outlook for Kenya because she believes young Kenyans will not tolerate government corruption as in the past.

## A SPECIES UNTO THEMSELVES

Although most attention now focuses on Richard, Meave, and Louise, some Leakey family members seem to retreat from the kind of publicity that made their surname so recognized in the past. Samira Leakey, the younger daughter of Richard and Meave, works for the World Bank in Washington, DC. Jonathan for many years owned a snake farm where he collected and extracted venom from poisonous snakes and trained snake handlers. He has also been involved in exporting pygeum bark from the *Prunus Africana* tree, used as an herbal treatment for enlargement of the male prostate.

After serving in Kenya's Parliament, Philip worked in ministerial positions with several governmental agencies. Recently he and his wife, Katy, have become involved in marketing environmentally friendly jewelry hand made by Maasai women, who have a long tradition of stunning bead work. The Maasai women make the jewelry from dead or renewable plants, woods, and dried grasses indigenous to Kenya. Through their connections, Philip and Katy have arranged to market the jewelry online and through an exclusive group of US stores, with money earned providing much-needed financial income to Maasai families.

Richard's half-brother Colin has spent his life in agricultural research, primarily at Cambridge University. His Internet homepage notes that he "is part of the Leakey clan of scientists who don't mind swimming against the tide until the tide turns in their favour" (http://www.colinleakey. com). A biologist with a doctor of philosophy degree, Colin's passion is nutrition and food, especially beans, to feed the world; "beans are good food value, high in protein and dietary fiber at a low price" (Ingrassia, 1997, A1). He is disappointed that the low-flatulent beans named Prim he has developed have not proved financially profitable. Colin feels large marketers of beans are reluctant to produce a gasless bean because it would send a negative message that their other beans produce gas.

Nevertheless he continues to toil on in the farm behind his home and preach the message that consumption of beans is a cheap, palatable way to improve world nutrition.

Louis's missionary family had core beliefs about working to better the world. In *Wildlife Wars*, Richard wrote that his mother referred to his own desire to contribute to Kenya as his "missionary genes" (p. 18), and family history confirms Louis and many of his descendents' choices as reflecting similar values regardless of the situation. Their story forms a pastiche of personalities, perseverance, and discoveries, and the world will be hearing from the Leakey family far into the twenty-first century.

# POSTSCRIPT

Writing a biography is such a personal experience that a writer often mentally becomes part of an extended community. My worldview grew immensely in writing *The Leakeys* and I not only mentally but also literally became part of the Kenyan community where the dramatic story of the first family of paleoanthropology takes place. I met, either through face-to-face or online communication, many people you read about in the preceding pages. I feel fortunate that the editors at Prometheus Books encouraged me to write this postscript and update you because, in the process, I satisfied my own curiosity about what has happened to some of the main characters since I initially researched and wrote this book.

Louise Leakey continues hunting for fossils with mother Meave. Professionally, they are moving ahead with plans for a five-year initiative in northern Kenya's Lake Turkana area, where their base camp is now located in Illeret. Louise has two daughters, Seiyia and Alexia. She also enjoys working in the family vineyard, which produces grapes for a fine white wine. She writes about this project at http://www .zabibu.org.

Richard Leakey speaks out on environmental issues and, with other well-known scientists, including Jane Goodall, is active with the Great Apes Survival Partnership (GRASP) http://www.unep.org/ grasp/, under the auspices of the United Nations. Because he believes that richer countries must contribute to funding in countries where most wildlife live, he and other conservationists initiated Wildlife Direct http://wildlife direct.org, a nonprofit conservation organization based in Kenya that uses the Internet to coordinate efforts and generate funding. He frequently lectures and is professor of anthropology at Stony Brook, a state university of New York, where he chairs the planning group for annual conferences on human evolution http://turkanabasin.org/.

Philip and Katy Leakey live in tents among the Maasai and are continually developing new jewelry and fashion accessories, with proceeds benefitting the Maasai who make them http://www.leakeylife.com/. Katy's blog is fascinating and rich with information about the Maasai women who make the Zulugrass and Zuluwood jewelry and other items that are available on their Web site and in stores across the United States.

Nasser Malit returned to Kenya in 2006 to collect data and visit Ethiopia to study *Homo erectus* fossils. With research grants from the Leakey Foundation and the American Museum of Natural History in New York City, he completed his collection of data, and anticipates soon receiving his doctoral degree from Binghamton University, State University of New York. His family now includes two children, with a son born in 2008.

Donald Johanson retired after twenty-seven years as director of the Institute of Human Origins (IHO) and has moved into the position of founding director; his good friend and colleague, Bill Kimbel, is the new director. Don enjoys traveling the world to give talks about Lucy and new advances in paleoanthropology and to raise funds for ongoing research and excavations by IHO scientists in Ethiopia and South Africa. His book *Lucy's Legacy: The Quest for Human Origins* was published in 2009.

I hope the fascinating story of the Leakeys and their search for our origins has inspired you to learn more about our human heritage.

Mary Bowman-Kruhm
writer@marybk.com
http://www.marybk.com
http://marybk.blogspot.com

# GLOSSARY

ACHEULIAN: A period of the early Stone Age, characterized by the making of stone tools, such as hand axes.

ANATOMIST: A specialist in the body structure of animals, especially humans.

ANTHROPOLOGY: The science made up of the fields of geology (rocks, soil, minerals, etc.), archaeology (ancient cultures), linguistics (language), and ethnology (comparison of cultures).

APE: Primate; chimpanzee, gorilla, or orangutan.

ARCHAEOLOGY: Scientific study of ancient cultures.

AUSTRALOPITHECINE: Prehistoric primate.

BCE: Abbreviation for Before the Common Era.

CE: Abbreviation for Common Era.

CHIMPANZEE: A small member of the ape family.

DERIVED FEATURES: Features of a fossil dissimilar to those in closest known ancestor.

DUIKER: A small African antelope whose short horns point backward.

ETHNOLOGY: Comparative study of different cultures.

GEOCHRONOLOGY: The science of determining the age of rocks and formations caused by geological events.

GEOLOGIST: One who studies the earth's crust with relation to rocks, soil, and minerals.

GORILLA: The largest ape.

GRACILE: Light, delicate build.

HOLOTYPE: A fossil that serves as the type specimen with which similar fossils are later compared.

HOMINID: A broad term for any primate that walks upright.

HOMININ: Today's humans and ancestors of today's humans; includes both humans and great apes.

HOMINOID: Ancestors of prehumans.

HOMO: Genus that includes human and extinct relatives.

HYRAX: A short-eared animal that weighs about five to nine pounds, eats vegetation, and is similar to a guinea pig or short-eared rabbit. It has hooflike toenails growing from the pad of each foot.

IN-SITU: In its original place; a fossil found *in-situ* would be found buried in sediment.

MATRIX: Soil or rock in which fossil is embedded.

OBSIDIAN: Glasslike material produced naturally by volcanic activity; used for tools because of the sharp blade it produces.

ORANGUTANG: A large ape found in Borneo and Sumatra.

OSTEOLOGY: Study of bones.

PALEOANTHROPOLOGY: The study of early humans and other primates based on fossil evidence.

PALEONTOLOGY: The study of how early humans lived based on fossil evidence.

PARADIGM: A conceptual framework or model that forms the basis of a particular theory.

PHYSIOLOGY: A branch of biology concerned with the internal working of living things, not size or shape.

PRIMATE: A member of the mammal family that includes humans, apes, and monkeys.

PRIMATOLOGY: Study of living nonhuman primates.

ROBUST: Heavily built body.

TUFF: Rock formed from layers of volcanic ash.

# BIBLIOGRAPHY

The following sources were used in synthesizing material presented in this book.

## WORKS ABOUT THE LEAKEYS

### Books and Chapters in Books

Bryson, B. 2003. *A short history of nearly everything.* New York: Broadway Books.

Cole, S. 1975. *Leakey's luck: The life of Louis Leakey, 1903–72.* London: Collins.

Darwin, C. 1979. *The illustrated origin of species.* New York: Hill & Wang.

Dibble, H. L., D. Williamson, and B. M. Evans. 2003. *Human evolution cookbook.* Philadelphia: University of Pennsylvania Museum of Archaeology & Anthropology.

Hamrum, C. I., ed. 1983. *Darwin's legacy.* New York: Harper & Row.

Hellman, H., 1998. *Great feuds in science.* New York: John Wiley & Sons.

Hunter, D. E., and P. Whitten, eds. 1978. *Readings in physical anthropology and archaeology.* New York: Harper & Row.

Isaac, G. L., and E. R. McCown, eds. 1976. *Human origins: Louis Leakey and the East African evidence.* Reading, MA: W. A. Benjamin.

Johanson, D. C., and M. A. Edey. 1981. *Lucy: The beginnings of humankind.* New York: Simon & Schuster.

Kalb, J. 2001. *Adventures in the bone trade.* New York: Copernicus.

Larson, G. 1989. *The prehistory of the far side.* Kansas City, MO: Andrews and McMeel.

Lewin, R. 1997. *Bones of contention: Controversies in the search for human origins.* Chicago: University of Chicago.

Lott, C., and B. Smith, eds. 2002. *Spectrum guide to Tanzania.* Brooklyn, NY: Interlink Books.

McGee, R. J., and R. L. Warms. 1996. *Anthropological theory: An introductory history.* Mountain View, CA: Mayfield.

McKee. J. K. 2000. *The riddled chain: Chance, coincidence, and chaos in human evolution.* New Brunswick, NJ: Rutgers University Press.

Morell, V. 1995. *Ancestral passions: The Leakey family and the quest for humankind's beginnings.* New York: Simon & Schuster.

Pike, J., ed. 2000. *Insight guide: Kenya.* London: APA Publications.

Posnansky, M., ed. 1966. *Prelude to East African history.* London: Oxford University.

Poynter, M. 1997. *The Leakeys: Uncovering the origins of humankind.* Berkeley Heights, NJ: Enslow.

Rasmussen, D. T., ed. 1993. *The origin and evolution of humans and humanness.* Boston: Jones and Bartlett.

Rowse, A. L. 1979. *The story of Britain.* New York: British Heritage Press.

Schwartz, J. H., and I. Tattersall. 2003. *The human fossil record.* Vol. 2. Hoboken, NJ: Wiley-Liss.

Tattersall, I. 2002. *The monkey in the mirror: Essays on the science of what makes us human.* New York: Harvest Book, Harcourt.

Tattersall, I., and J. H. Schwartz. 2001. *Extinct humans.* Boulder, CO: Westview Press.

Weitzman, D. 1994. *Human culture: Great lives.* New York: Charles Scribner's Sons.

Willis, D. 1992. *The Leakey family: Leaders in the search for human origins.* New York: Facts On File.

## Articles

Academy of Achievement. 1991. Interview with Donald Johanson. January 25, 1991. http://www.achievement.org/autodoc/page/joh1int-6 (accessed November 12, 2004).

Barsby, J. 2004. The alternative: Kenya wildlife service collection. *Travel News & Lifestyle: East Africa* (March): 24–25.

Begley, S. 1984. *Newsweek*, October 29, 104, 119.

Bellers, V. 2001. What Mr. Sanders really did or a speck in the ocean of time (chapter 18). (Unpublished manuscript). http://www.britishempire.co.uk/ article/sanders/sanderschapter18.htm (accessed November 26, 2004}.

Boam, Jeff. 1989. *Indiana Jones and the Last Crusade*. http://www.corky.net/ scripts/indieAndTheLastCrusade.html (accessed November 26, 2004).

Bunn, H. T. 1991. A taphonomic perspective on the archaeology of human origins. *Annual Review of Anthropology* 20: 433–67.

Clark, C. M. 1989. Louis Leakey as ethnographer. *Canadian Journal of African Studies* 23: 380–98.

DeSilva, J. 2004. Interpreting evidence: An approach to teaching human evolution in the classroom. *American Biology Teacher* 66 (April): 257–66.

Eisner, R. 1991. Scientists roam the habitat as zoos alter their mission [Electronic version]. *Scientist* 5, no, 11 (May 27): 1–5. http://www. the-scientist .com/yr1991/may/ eisner_p1_910527.html.

Fleagle, J. G. 2004. Centennial tribute to L.S.B. Leakey, *Evolutionary Anthropology* 13: 3–4.

Gore, R. 1997. The dawn of humans: Expanding worlds. *National Geographic* (May): 84–109.

Halliday, T. 1996. Review of *Science masters: The origin of humankind*. *Journal of Biological Education* 30, no. 2 (June 1): 143.

Hardin, B. 2000. The last safari. *New York Times Magazine*, June 1, 2000, 66 70.

Hunsinger, A. 1997. Remembering Mary Leakey. *AnthroQuest* 3 (Spring): 1–3.

Ingrassia, L. 1997. Dr. Colin Leakey, a real bean counter finds profit elusive. *Wall Street Journal*, April 1, 1997, A1, A12.

Ippolito, M. 2004. Exhibit: What Goodall taught us about chimps. [Electronic version]. *Atlanta Journal-Constitution*, February 25, 2004. http://www.ajc .com/metro/content/metro/atlanta/ 0204a/ 26chimp.html (accessed June 6, 2005).

Jablonski, N. 2004. Putting technology to work at Koobi Fora. http://www.kfrp.com/dispatches_2004/gis_jablonski/gis_jablonski.htm (accessed May 31, 2005).

Johanson, D. C. 1999. The Leakey family. *Time* 153, March 29, 180–83.

Kirby, A. 2004. Fences "can help apes' survival." *BBC News Online,* May 5, 2004. http://news.bbc.co.uk/1/hi/sci/tech/3686783.stm (accessed June 29, 2004).

Lacey, M. 2003. The Saturday profile: A scholar follows her family's dusty footprints. [Electronic version]. *New York Times,* April 19, 2003, A-4.

Lewin, R. 2002. The old man of Olduvai Gorge. *Smithsonian* (October): 82–88.

Lieberman, D. E. 2001. Another face in our family tree. *Nature* (March 22): 410, 419–20.

Morell, V. 1996. The most dangerous game. *New York Times Magazine,* January 7, 1996, 31–33.

Nalencz, V. K. 2005. Stones & bones at the dawn of humanity. *Temple Review* 58, no. 2 (Spring): 24–29.

Out of Africa. 2004. *Interior Design* (April): 96.

Rennie, J., ed. 2003. New look at human evolution [Special Edition]. *Scientific American.*

Richard Leakey. 1994. *Newsmakers 1994,* 4. Gale Research, 1994. Reproduced in Biography Resource Center. Farmington Hills, MI: Gale Group, 2004. http://galenet.galegroup.com/serlet/BioRC.

Scout report. 2004. Noted environmentalist urges immediate action to save world's great apes. May 7, 2004. http://scout.wisc.edu/Reports/ScoutReport/ 2004/scout-040507.inthenews (accessed June 29, 2004).

Simons, A. J., and Z. Tchoundjeu. 1998. Passing problems: Prostate and *Prunus;* African team works to maintain sustainable supply of pygeum bark. [Electronic version]. *HerbalGram* 43: 49–53. http://www.herbalgram.org/youngliving/herbalgram/articleview.asp?a=1056 (accessed November 28, 2004).

Skeletons in the family closet. 1999. *Discover* 20 (May): 5. http://firstsearch.ocle.org/ (accessed February 16, 2003).

Smithsonian Institution Human Origins Program. *Paranthropus aethiopicus.* http://www.mnh.si.edu/anthro/humanorigins/ha/aeth.html (accessed November 10, 2004).

Strahle, J. 2001. Scholar lives up to family name. [Electronic version]. *New*

*Media Index, Truman State University* 92 (April 5): 25. http://index
.truman.edu (accessed March 14, 2003).

University of Utah Media Release. 2001. Skull of new early human relative
found in Kenya. March 21, 2001. http://www.newswise.com/p/articles/
view/23061 (accessed June 6, 2004).

Van Couvering, J. 2001. Review of *Adventures in the bone trade: The race to
discover human ancestors in Ethiopia's Afar depression.* [Electronic ver-
sion]. *Natural History* (June).

White, T. 2003. Early hominids—diversity or distortion. *Science* 299 (March
28): 1994–97.

## Electronic Media

http://www.amonline.net.au/human_evolution. Web site of the Australian
Museum Online. Clear and concise introduction to human evolution.

http://www.antiquityofman.com. Web site with three focus areas: Ancient
Egypt, hominin evolution, and pseudoscience; hosted by Mikey Brass,
African archaeologist.

http://www.archaeologyinfo.com/. Excellent information and extensive glossary from anthropologists who developed this Web site as "a forum for the
unity of ideas and synthesis of a common human evolutionary model."

http://www.becominghuman.org/. Web site of the Institute of Human Origins; a
variety of resources, including a glossary, list of relevant Web sites sorted by
topic, and a broadband interactive documentary hosted by Donald Johanson.

http://www.globalmarketstore.com/zulnectan.html. Web site that sells jewelry
hand made by Maasai women.

http://globetrotter.berkeley.edu/people3/White/white-con0.html. Conversa-
tions with History; Institute of International Studies, University of California Berkeley. Interview with Dr. Tim White.

http://www.leakeyfoundation.org/. Web site of the foundation established to
support research into human origins.

http://www.madsci.org. Search for *hominin* and *hominid* to retrieve a post that
traces the evolution of both terms in everyday language.

http://www.mnh.si.edu/anthro/humanorigins. The Smithsonian Institution
Human Origins Program. Includes Dr. Rick Potts's fieldwork dispatches
from Olorgesailie.

http://www.mos.org/evolution. A Web site useful for the classroom, devoted to an understanding of evolution, with up-to-date family trees representing three different interpretations of the fossil record and a unit for teaching human evolution in the classroom.

http://www.museums.or.ke. National Museums of Kenya Web site.

http://www.npr.org/programs/re/archivesdate/2003/aug/leakey/index.html. Interviews with family members to commemorate Louis Leakey's 100th birthday.

http://www.tradgames.org.uk/games/Mancala.htm. Information about game of bau found in *Online Guide to Traditional Games: History and Useful Information.*

## Other Media

Eddings, J. (interviewer and managing editor). 2003. Interview with Louise Leakey. [Television interview]. In series *On Africa*. Washington, DC: WHUT Howard University Television.

Suarez, R. (interviewer). 2002. Interview with Meave Leakey. [Television interview]. In Online NewsHour: New Beginnings. A NewsHour with Jim Lehrer Transcript. March 26, 2002. http://www.pbs.org/newshour/bb/science/jan-june01/newbegin_03–26.html (accessed November 29, 2004).

## WORKS BY THE LEAKEYS

## By Louis Leakey

Leakey, L.S.B. 1954. *Defeating Mau Mau*. London: Methuen & Co.

———. 1966a. *Kenya: Contrasts and problems*. Cambridge, MA: Schenkman Publishing. (Original work published 1937.)

———. 1966b. *White African: An early autobiography*. New York: Ballantine Books. (Original work published 1937.)

———. 1970. *Stone age Africa: An outline of prehistory in Africa*. New York: Negro Universities Press. (Original work published 1936.)

———. 1974. *By the evidence: Memoirs, 1932–1951*. New York: Harcourt Brace Javanovich.

## By Mary Leakey

Leakey, M. 1983. *Africa's vanishing art.* Garden City, NY: Doubleday.
————. 1984. *Disclosing the past: An autobiography.* New York: McGraw-Hill.

## By Richard Leakey

Leakey, R. E. 1986. *One life: An autobiography.* Salem, NH: Salem House.
————. 1994. *The origin of humankind.* New York: Basic Books.
Leakey, R. E., and R. Lewin. 1977. *Origins.* New York: E. P. Dutton.
————. 1993. *Origins reconsidered: In search of what makes us human.* New York: Anchor Books.
————. 1995. *The sixth extinction: Patterns of life and the future of humankind.* New York: Doubleday.
Leakey, R. E., and V. Morell, 2001. *Wildlife wars: My fight to save Africa's natural treasures.* New York: St. Martin's Press.
Walker, A., and R. E. Leakey, eds. 1993. *The Nariokotome* Homo erectus *skeleton.* Cambridge, MA: Harvard University Press.

## By Louise Leakey

Leakey, L. 2004a. KFRP field season dispatches: Week 3 (February 2004). http://www.kfrp.com/dispatches_2004/dispatch02/2004dispatch02.htm.
————. 2004b. KFRP field season dispatches: Week 4 (February 2004). http://www.kfrp.com/dispatches_2004/dispatch03/2004 dispatch03.htm.
————. 2004c. KFRP field season dispatches: Week 5 (February 2004). http://www.kfrp.com/dispatches_2004/ dispatch05/2004 dispatch05.htm.
————. 2004d. KFRP field season dispatches. Week 6 (March 2004). http://www.kfrp.com/dispatches_2004/dispatch06/2004 dispatch06.htm.
————. 2004e. KFRP field season dispatches: Week 11 (May 2004). http://www.kfrp.com/dispatches_2004/dispatch11/2004 dispatch11.htm.
————. 2004f. KFRP field season dispatches: Week 12 (May 2004). http://www.kfrp.com/dispatches_2004/dispatch12/2004 dispatch12.htm.
————. 2005. KFRP field season dispatches: Week 4. http:// www.kfrp.com/ dispatches_2005/dispatch04/ dispatches_2005_04.htm.

## Electronic Media

http://www.colinleakey.com Web site of Louis's son, an agricultural researcher at Cambridge University.

http://www.kfrp.com Web site describing the Koobi Fora Research Project with dispatches by Dr. Louise Leakey; also includes information about the Samburu hominin excavation.

http://www.leakey.com The Leakey family Web site.

## Other Media

Leakey, R. E., speaker. 2003. Richard reflects upon the challenge and excitement of his four careers, August 7. National Public Radio. Retrieved June 6, 2004, from http://www.npr.org/ programs/re/archivesdate/2003/ aug/leakey/ index.html.

## Speeches

Leakey, L. 2003. Presentation at the National Geographic Society, Washington, DC, October 16.

Leakey, M. E., and L. Leakey. 2003. *Discovering Our Earliest Ancestors*. Presentation at the Louis Leakey Centennial Celebration, Field Museum, Chicago, IL, October 10.

# INDEX

# ABOUT THE AUTHOR

**M**ARY BOWMAN-KRUHM is author of over thirty books for children and young adults and a Faculty Associate, Johns Hopkins University, College of Education. Web site: http://www.marybk.com